Harold Hill's lunar portfolio is a unique collection of drawings published for the first time in this form. Each illustration is supplemented with descriptive text, usually based on field notes taken at the time of observation. Astronomical drawing still has an important place alongside photography. Indeed, drawings constructed by an artist who takes advantage of the fleeting moments of perfect vision are often more detailed than photographs.

A portfolio of lunar drawings

The Practical Astronomy Handbook series

The Practical Astronomy Handbooks are a new concept in publishing for amateur astronomy. These books are for active amateurs who want to get the very best out of their telescopes and who want to make productive observations and new discoveries. The emphasis is strongly practical: what equipment is needed, how to use it, what to observe, and how to record observations in a way that will be useful to others. Each title in the series will be devoted either to the techniques used for a particular class of object, for example observing the Moon or variable stars, or to the application of a technique, for example the use of new detectors, to amateur astronomy in general. The series will develop into an indispensable library of practical information for all active observers.

THE CENTRAL MT. GROUP & GREAT RILLE

OF **PETAVIUS**

AT SUNSET

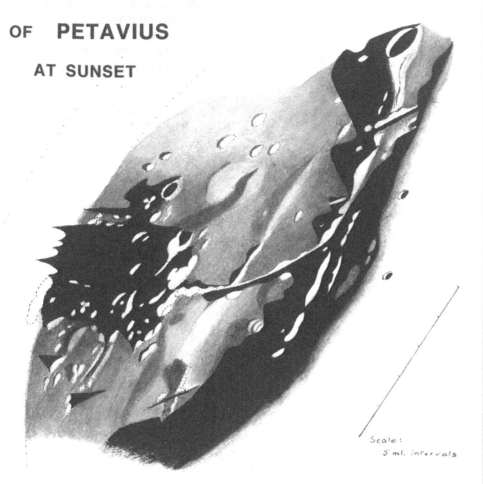

Scale:
5 ml. intervals

Geoc. Libr.   { L = +4°.09
for 00ʰ   { B = -6°.56
              (v. favourable)

1989 October 17
00ʰ00ᵐ – 01ʰ30ᵐ U.T.

10" Reflector ×286
Seeing variable: 4 ~ 7 or 8/10
Transparency 3 – 4/5

⊙'s { Colong. 116°.44 – 117°.20
    { Sel. Lat. –1°.32

L.N. 826

A portfolio of
lunar drawings

Harold Hill

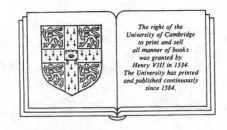

CAMBRIDGE UNIVERSITY PRESS

Cambridge
New York Port Chester
Melbourne Sydney

PUBLISHED BY THE PRESS SYNDICATE OF THE UNIVERSITY OF CAMBRIDGE
The Pitt Building, Trumpington Street, Cambridge, United Kingdom

CAMBRIDGE UNIVERSITY PRESS
The Edinburgh Building, Cambridge CB2 2RU, UK
40 West 20th Street, New York NY 10011–4211, USA
477 Williamstown Road, Port Melbourne, VIC 3207, Australia
Ruiz de Alarcón 13, 28014 Madrid, Spain
Dock House, The Waterfront, Cape Town 8001, South Africa

http://www.cambridge.org

First published 1991
First paperback edition 2003

A catalogue record for this book is available from the British Library

Library of Congress cataloguing data available

ISBN 0 521 38113 4 hardback
ISBN 0 521 54208 1 paperback

This portfolio is dedicated in gratitude

to the memory of Alfred Noël Neate, one of the earliest members of the British Astronomical Association. He was my mentor in lunar studies in my early observing days, stressing the paramount importance of accurate positional work in selenography. Neate's measures ranked with those of Franz and Saunder. A man of integrity and high accomplishment,

to Richard M. Baum, eminent in his field in Astronomy, without whose drive, unquenchable enthusiasm, and continual encouragement, this book would certainly never have appeared,

and finally, above all, to my long-suffering wife, Phyllis, truly an astronomer's widow in many respects, who has indulged me uncomplainingly in my long hours both at the telescope and in my study for so many years.

Many a pleasant hour awaits the student in these wonderful regions; only let him not expect that what he sees so plainly will be equally intelligible, except in its unquestionable relief from the effects of light and shade

Rev. T. W. Webb (1806–1885)

Contents

Foreword

To categorise Harold Hill as an amateur astronomer is too bland. It conveys nothing of the man who, for almost half a century, weather and leisure permitting, has assiduously charted the play of light and shade across the surface of the moon in a series of exquisite drawings. It tells even less of his eager anticipation and how his published work has encouraged and inspired moonwatchers the world over. Nor does it inform us of his approach to selenography in this age of moon landings and computer-aided telescopes. The fact is, this modest, almost self-effacing man is something of a legend. He is a living link with the golden days of moonwatching; one of the few remaining practitioners in the tradition begun by Galileo and Thomas Harriot, that includes such names as J. H. Schröter, W. G. Lohrmann, Beer and Mädler, J. F. J. Schmidt, E. Neison, T. Gwyn Elger, W. R. Birt, P. J. H. Fauth, J. N. Krieger, L. Weinek, W. Goodacre and H. P. Wilkins.

All very well it may be said. But what, other than for personal satisfaction and self-education, is the purpose of the enterprise? Why should someone spend long hours, often in great discomfort, drawing what has been photographed in great detail many times before, and at close range? So expressed, the activity assumes a certain irrelevance. And yet it has form and meaning. For what differentiates the artist from the camera, is the former's ability to discern a mood, a brief moment of truth, in its full context. The event may be transitory, unimportant even, when seen in isolation, but its occurrence transforms the ordinary and invests it with a quality that illuminates the obscure and provides new insights into old knowledge. Here on earth we commonly observe the weather. Some will comment on the excessive dampness. Others, the strength of the wind. Still others, the vagaries of temperature. But who of all those vociferous commentators will tell us of the shafts of golden light which stabbed out of the western cloud late yesterday? Or the glowing cloudscape at sunset? All passed unremarked.

So it is with moonwatching. Many look, few observe. Optical aid, no matter how slight, opens the grandeur of the scene to everyone, given the will to look. Sunrise on the Apennines with its gorgeous chiaroscuro effects. The stark beauty of the Mare Crisium under a low sun. Rugged profiles along the southern limb. The enchanting delicacy of the Triesnecker rille system. These are the haunts of the casual observer who in the showpiece perceives a vehicle for therapeutic relaxation, and derives an intangible benefit from metaphysical contact with another world. But the truth is not always to be found in the dramatic as John Ruskin observed, but, '. . . in the quiet and subdued passages of unobtrusive majesty, the deep, the calm, and the perpetual . . . which are to be found always yet each found but once.'

This is where Hill sparkles. He long ago abandoned the familiar. An admission which prompted him to confess: 'Even now, after so many years, I feel my topographical knowledge of the moon is desperately limited.' If only it were! Still I understand what he means. It is the Chinese box syndrome. The realisation that comes with experience, or in those rare moments when there seems to be a hole in the atmosphere.

I had my first sight of a Harold Hill moon drawing in the late 1940s and was impressed. Now the master has an unrivalled knowledge of the lunar terrain, and that first impression has crystallized into outright admiration for the accuracy and thoroughness of his descriptions. Completed in stipple and wash, his artwork is bright and inimitable. It has attracted many imitators, but few have the dedication of its creator, or his eye for detail, fewer still his artistry. Qualities that have brought Hill professional recognition. And still he refines his technique. For each night, when the sky is clear and the moon is above the horizon, he is to be found at the eyepiece of his telescope, observing and drawing with undiminished enthusiasm.

Such tireless devotion may puzzle a generation brought up on Apollo imagery. Hill too has questioned his motives. To ask why of him is to invite a typical response. 'I observe because I like it. Isn't that enough?' Of course it is. Nevertheless it is unsatisfactory. It leaves me with the feeling it is all too easy. There is no simple answer, for all is mixed; individuals refuse to be neatly classified and catalogued. Be that as it may, in the words and imagery of those who embark on out of the ordinary ventures, a picture emerges which is curiously informative of the underlying metaphysics. By a myriad ways we probe the unknown, as we seek for a new and fuller experience of life. A tiny boat on a storm-tossed sea. A modest telescope aimed at the stars. Different in kind, but united in a common pursuit by the self-same spirit which from time immemorial has 'called with strange insistence to him who, wondering on the world, felt adventure in his veins', as Percival Lowell wrote in 1906. A sentiment echoed by William Herschel who resolved, '. . . to take nothing on trust, but to see with my own eyes all that other men had seen before.'

Harold Hill himself gave vent to this expression in April 1988 when he wrote to say he had just witnessed, '. . . the last lighting on the great E. wall of Clavius,' adding, 'that I have waited *years* to catch this right.' He went on to say, 'I still await an opportunity to follow the very last stages of illumination to see if the upper crests break into isolated peaks.' Six months later, five beautifully figured drawings told of success. Made during the early hours of a chill November morning, they record in vivid detail the creep of shadow up magnificent ramparts, until finally only the tops are left glistening star-like above slopes deeply involved in night. On the floor, far below, a number of crater-pits bask in a pool of shadowy light. The scene is moody and atmospheric. It reminds us that Harold Hill is not just a mapper of dead rocks. He looks at the moon with the eye of an artist. His moon is indeed that of Galileo: '. . . uneven, rough, and full of cavities and prominences . . . relieved by chains of mountains and deep valleys.' But it is also the moon of the elder Herschel with its erupting volcanoes and dense forests; the enigmatic walled city of Gruithuisen, along with the industrial smogs of J. H. Schröter. 'This charming moon,' observed Camille Flammarion, 'has undergone in human opinion the vicissitudes of this opinion itself, as if it had been a political personage.' The art of Harold Hill has the quality to remind us of this fact. For, in the shadows and highlights, we perceive something more than harsh reality. An indefinable something, caught on the edge of things, that is as revelatory of the man as the moonworld itself. Intangible it may be, but it cloaks the scars of a violent past with a mystique as telling as any physical discovery, reminding us of the excitement which awaits those who embark upon a personal journey of discovery and exploration. It also reminds us not to overlook the aesthetic, philosophical and spiritual aspects of observing. After all, these may have the greatest value to the individual. Moreover, to observe the moon in all its phases is to be made aware of its place in the human psyche. But, to furnish the mind with the refined pleasure thus obtained, it is necessary to look for yourself. This is the measure of Harold Hill's singular achievement.

RICHARD BAUM
Chester
February 1990

Introduction

This is not a guidebook to the Moon in the accepted sense, nor is it an instruction manual. It is, in effect, and as the title implies, a collection of drawings representative of various parts of the lunar surface and illustrative of different aspects of investigation which have been selected from the author's observing files covering more than 40 years.

In some respects, the choice of subjects depicted may not satisfy every reader and some may be disappointed that familiar objects such as Aristarchus, Copernicus, Plato and Theophilus, to mention a few, do not feature in these pages. This is because formations such as these were singled out for special study by earlier generations of observers. It has long been the present writer's contention that many equally deserving regions have not received anything like the same attention; accordingly he decided that some of these comparatively neglected areas should be included as warranting further study.

Not unnaturally, the present selection of some 95 regions represents features of especial interest to me, and the reader will find greater elaboration of the reasons for their selection in the notes which accompany the drawings. Suffice it to say at this point, that an early realisation that much remained to be done in mapping these wonderful landscapes led me to undertake serious lunar study almost from the commencement of my amateur observing. As for the wide gaps which existed in our cartographical knowledge of the surface, it is only fair to point out that the lunar globe is a truly vast place for telescopic exploration and as H. P. Wilkins, the selenographer once remarked in personal correspondence, 'No man, even if he were to live to the age of Methuselah, could possibly observe effectively all the lunar formations and make a complete map of the Moon from his work alone'. This was written at a time when exploration was made with cartography very much to the fore in traditional continuation of the earlier map-makers such as Mayer, Schröter, Lohrmann, Schmidt, Neison, Elger and Goodacre.

In the second half of the present century, however, with the advent of much improved techniques in the form of terrestrially based photography and high-resolution records obtained from the space missions, in which the photographic coverage of *almost* the whole surface was accomplished, it might be supposed that traditional selenography was all but redundant. The impression could have been given that a revolution had taken place in which the amateur's role, formerly regarded as of the greatest importance, had now given way to a massive professional involvement.

To the contrary, there is still much useful work for the Earth-based visual observer to do, even though one labours under the disadvantage of a more or less fixed viewpoint in space, positioned, as one must inevitably be, at the bottom of a usually very turbulent ocean of atmosphere. It cannot be over-emphasised that, despite the thousands of close-range photographs we now have of the Moon, a fair percentage of these are limited to certain aspects of lighting only and are what may be loosely termed, 'one-offs'. Indeed, in seeking their aid, say, in settling some observational query, one is all too often thwarted by inappropriate lighting or, perhaps, unsuitable positioning of the space camera at the time of exposure.

Amateurs should not allow themselves to be deterred in their efforts by the excellence of professional results. As in so many fields of human endeavour, the qualities of patience, application, assiduity – all summed up in the word dedication – when coupled to experience, skill and, of course, suitable equipment, can still achieve a good deal even though the emphasis and direction of the observer's studies may have had to change.

Quite apart from these considerations, there is another aspect to the business of observing which must not be disregarded, namely, the sheer aesthetic pleasure to be derived from contemplation of the stark grandeur of the lunar landscape under the ever-changing solar illumination. Many of the drawings in this book were inspired by a compulsion to attempt a representation of what was before me rather than from any scientific intent – this is indicative of the glorious freedom of the mere amateur who, unlike the professional, is not constrained to work under controlled, disciplined procedures except, of course, where such strictures are self-imposed. It is frequently asked why amateurs continue to engage in pursuit of 'such aimless observations', the instant rejoinder to this may best be expressed in the Latin tag – 'Labor Ipse Voluptas'.

Professor George Ellery Hale's definition of the true amateur is as appropriate today as it ever was:

'a man who works in astronomy because he cannot help it, because he would rather do such work than anything else in the world and therefore cares little for hampering conditions of any kind'

The number of such people can never be too large and it is the prime object of this portfolio to stimulate and encourage those who are drawn to the study of our nearest neighbour in space, but *not* those whose preference is to do so only from the comfort of an armchair.

The preparation of lunar drawings

On many occasions over the years I have been asked for advice and information on the best method of making lunar drawings and the stages leading to the 'finished' product. It is hoped that the following notes may prove of assistance to others.

Equipment

It goes almost without saying that making an effective drawing at the telescope is a fairly difficult exercise and a very different matter from confronting, say, a still-life subject in the studio. The situation is incomparably more difficult because we have a continually moving target *unless* the telescope is equatorially mounted and provided with an efficient drive – the latter has the obvious advantage of leaving both hands free to hold the board and make the drawing. Regarding the drawing equipment, I have found nothing to improve on a piece of hardboard or stout plywood of adequate size (11″ × 13″) with one or more bulldog clips to secure the sheet of cartridge paper. A permanent attachment at the top of my board is a 6″ length of strip lighting – the intensity of which can be controlled by rheostat to suit the observing conditions; this is fed from the mains supply and shaded so that the illumination falls upon the paper only, thus preventing direct glare to the eye. Some observers prefer a more portable battery arrangement because of its lightness and freedom from a trailing cable but, in my opinion, the advantage of the former method outweighs these factors. It is important to adopt the most comfortable position possible at the eyepiece, as working under any form of constraint will lead inevitably to poor results.

Drawing

My own practice has always been to make careful *line* sketches employing an intensity scale from 0 –

indicating black shadow (actually the letter S is substituted as 0 could be misconstrued in the analysis as a marking!) to 10 for the most brilliant parts – usually taken as the brightness of the central peak in Aristarchus. A fairly reliable calibration for intermediate intensities is possible if one adopts 4 for the intensity of maria some 3 days from the terminator although this can vary dependent upon which mare is chosen. Basically, the scale used is that formulated by Schröter in the late eighteenth century and later adopted by Beer & Mädler and then by Elger. It is useful to commit the basic scale to memory or to have it within easy reference – particularly its further elaboration; see page xxii, where both are given. Fractional estimates of a tonal degree are not always easy to make and generally I do not go much below $\frac{1}{2}$ unit – even so, results must be regarded only as relative, not absolute. Nevertheless, this method may be considered as quite adequate for the purpose of making a reliable *toned* drawing from the observational data. One advantage of this system is that, provided the original drawing is conscientiously done, it is possible to prepare a satisfactory 'finished' representation of what was seen two days, two weeks or even two years on without having to rely on that notoriously uncertain faculty – memory. See stages (1) and (2) of the drawing of Gassendi on page xv. It should hardly be necessary to add that all drawings should be labelled with the date, time (in Universal Time), telescope type and aperture, magnification, quality of the seeing, transparency, etc.

Perhaps a word about the scale to be adopted may not come amiss – much will depend on the image size and its quality. These factors are dependent on the telescope's aperture, its excellence, the magnification used and, above all, the quality of seeing. All these parameters are necessarily interlinked. On the whole, I find a scale of 15–20 miles to the inch a comfortable size as it

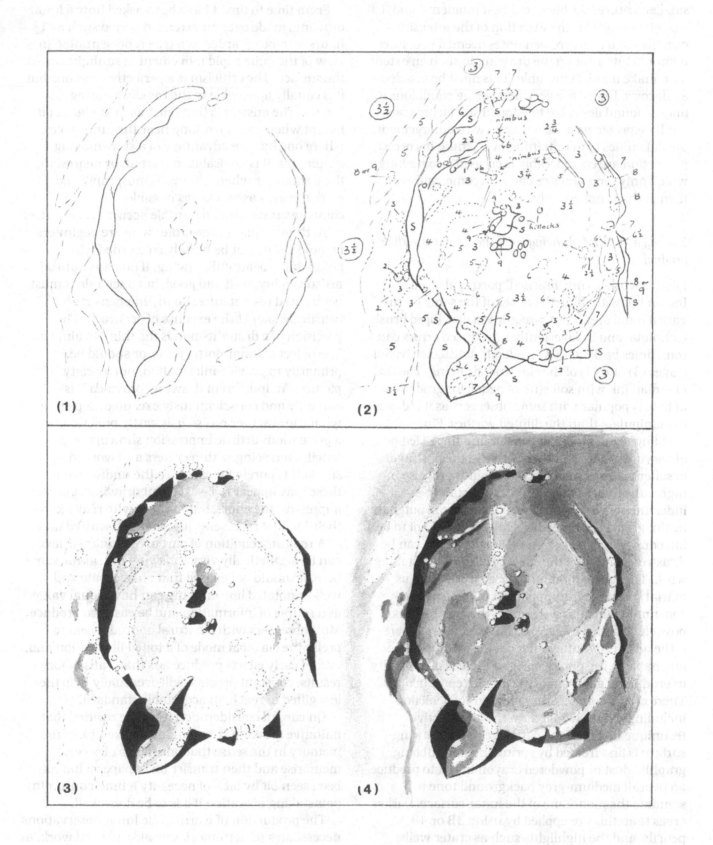

Stages in the preparation of a lunar drawing

enables one to depict clearly the true nature of the smaller features visible in the best moments, and it also gives space for the insertion of the intensity numbers and any arrowed notes deemed necessary without cluttering up the drawing to such an extent as to make it indecipherable. This must be avoided at all costs. Under the very best seeing conditions it may be found desirable to adopt a still larger scale but here problems arise because, with enlargement, the difficulties of maintaining everything in correct proportion are considerably increased – something which only long practice can overcome, and even then perhaps not entirely.

Making a 'finished' drawing from the telescope outline original

I like to work up my 'finished' portrayals using Indian ink with diluted washes of the same for the varied tonal areas, but it requires a developed brush technique and inadvertently introduced errors can sometimes be difficult to eradicate satisfactorily. See stages (3) and (4) of the Gassendi drawing. The use of Indian ink with soft stub of pencil for gradations of tone is popular with some observers as it is easier to manipulate than the diluted washes. Fine stippling using mapping pens or fine-tipped felt pens of varying grades can be effective in the portrayal of softer features such as crater-bands and also in high-sun studies where there is a certain indefiniteness in the outline of lunar detail but, here again, great care is needed if false detail is not to be introduced and, unless one is adept, results can be disastrous and spoil the effect of a lot of painstaking work. The representation of large tracts such as marial surface using stippling is obviously time-consuming and also a strain on the eyes but it is possible to achieve very pleasing results and there is the added advantage that such a medium is amenable to the photocopier so there is no difficulty in producing good-quality multiple reproductions. There are, of course, other methods of depiction including what I would term the 'subtractive' technique in which the cartridge paper drawing surface is first treated by sprinkling and rubbing graphite dust or powdered crayon into it to produce an overall medium-grey background tone to simulate the monotone of the lunar surface. Darker areas than this are applied by using 3B or 4B pencils, and the highlights such as crater walls, sunlit elevations and the like, are picked out using hard, finely chamfered erasers – all in all a pretty messy business with the end result far from convincing . . . at least in the examples which the author has seen. This is a method which he has never even considered adopting.

From time to time I have been asked how a lunar drawing, made over an extended period such as $1\frac{1}{2}$ hours, can be regarded as a true representation in view of the quite rapid movement of sunlight across the surface. The criticism is a perfectly valid one but it is equally applicable to the rapidly rotating Jupiter. The answer is that one has to 'dodge' a bit. Except when one is working near the lunar poles, where one has the advantage of a slow-moving terminator, it is advisable to start at, or nearest to, the terminator where change is most rapid and work inwards as quickly as possible – commensurate with all possible accuracy, of course.

To those reading these notes who are beginners I would say do not be unduly concerned with producing a beautiful drawing. If one has natural artistic ability, well and good, but such talent must *not* be used (as not infrequently happens with artistic people) at the expense of accuracy. The precision of a draughtsman is the thing to aim for. The object of lunar portrayal is, or should be, primarily to provide information, not a pretty picture. An indifferent drawing, provided it is carefully and conscientiously executed, depicting what the observer has *seen*, is vastly preferable to a pretentious artistic impression showing, say, fanciful terracing within craters and worked-up surrounds purely for effect. To the undiscerning these may appear to be excellent drawings and very impressive to behold, but to those who really know their Moon it represents just so much wasted time.

A realistic rendition of part of the lunar surface can be aesthetically very satisfying but, at the same time, it should be realised that an accurate and well-annotated line drawing can have equal value as a source of information and be easier to produce. Most observers with a natural aptitude seem to prefer the classical mode of a toned illustration and, even if early efforts produce less than satisfactory results, constant practice will eventually help the less gifted to reach an acceptable standard.

On careful consideration it will be realised that imitative drawing is essentially an exercise of the memory in the sense that one has to observe, memorise and then transfer on to paper what has been seen bit by bit – of necessity a time-consuming, painstaking operation if it is to be done well.

The production of worthwhile lunar observations necessitates something closely akin to hard work, as the concentration and co-ordination of hand and eye required, especially over long periods of perhaps difficult seeing and trying weather conditions, is, in my opinion, every bit as arduous as actual physical toil. Only those of us who have crept off to bed after a long vigil at the telescope, tired and perhaps stiff with cold (in the winter months) can fully appreciate the truth of what has just been written!

Another question often posed regarding drawing at the telescope is whether it is justifiable to prepare outlines of formations from photographs as a means of saving valuable time at the eyepiece and, at the same time, ensuring some degree of accuracy with respect to the shape and position of the more salient features.

On principle, I am not in favour of this procedure mainly because I feel it would hinder or inhibit the development of drawing ability by beginners if habitually followed. Whilst conceding that it has merit as a time-saver, it is applicable only to the central regions of the disc which suffer little from the effects of libration (one assumes here that most readers will be conversant with such fundamental terminology as applied to the Moon). It is certainly not a practice to be entertained in the case of formations lying 40° selenographic or more from disc centre where considerable distortion can occur from libratory swing and render prepared outline sketches useless. A striking and classic case is that

of Gassendi which, though occupying only a borderline position on 40° can be subject to gross alteration in shape, made more noticeable on account of its markedly polygonal outline.

To reinforce what has been written about changes in form and the extent of displacement from the mean position caused by libration, it is axiomatic that such effects will be accentuated to an increasing degree as the limb is approached. In the late 1950s a team of observers was contributing observations of the Miyamori Valley region (between Lohrmann and Riccioli) to the B.A.A. Lunar Section – it was a subject which attracted much attention at that time from lunarians. However, some puzzlement and concern was expressed about the gross inconsistencies in the directional line of the Valley as represented on published drawings of different dates. I pointed out that the apparent errors were largely, if not entirely, explained by the effects of libratory movement in latitude – values which were in constant change; furthermore, the closer objects were carried towards the limb by the additional motion of libration in longitude, the greater the apparent displacement of features relative to each other especially when

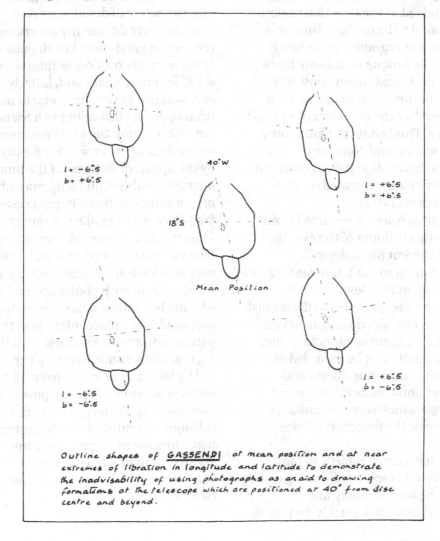

Outline shapes of GASSENDI at mean position and at near extremes of libration in longitude and latitude to demonstrate the inadvisability of using photographs as an aid to drawing formations at the telescope which are positioned at 40° from disc centre and beyond.

compared under opposing northern and southern tilts. To make this clear I prepared three charts for publication which demonstrated better than any attempted verbal explanation how these effects were brought about and they are here reproduced.

Note the very considerable change in the *direction* of the axis of the Miyamori Valley (marked heavily in black) with respect to a line radial to the nearest point of the limb. The extent of the shift in position of, say, the northern border of Lohrmann with respect to the northern limit of Riccioli is also worthy of note (see page xix).

From the foregoing remarks it will be evident, therefore, that only in very exceptional circumstances is the use of prepared outlines from photographs warranted for observations in the zones beyond 40°, otherwise the inexperienced observer will be confronted with some puzzling differences when he goes to his telescope.

Some thoughts on the art of vision

There can be no doubt that an experienced observer sees much more detail on the Moon than a beginner and this fact is the result of the long training of the eye. The advantage of this experience has not, to my knowledge been denied or even seriously challenged by anyone in a position to judge. There are instances where an object, initially perceived with difficulty is, at some later date, seen more easily with the same instrument under virtually identical observing conditions. Even more cause for surprise is that a feature seen in a moderately sized telescope, say a 10-inch, and judged at the time to be near its resolution limit, is found to be readily visible in a much smaller instrument *providing its position is known*. Of course, the technique of utilising those moments when the air is most steady, however fleeting, to grasp detail concealed in mediocre or poor seeing is employed by the experienced to build up the overall picture. This is how the lunar (and planetary) observer works, and continued application improves this facility. It is, in effect, an exercise which, by active concentrated effort, develops and enhances visual skill.

It was quite some time before it occurred to me that there is more to this training of the eye: the characteristic of its inherent physiological adaptation or conditioning to the brightness level of the telescopic image whereby it learns (unconsciously to its owner, perhaps) to distinguish ever so delicate contrasts of albedo or tone on an extended surface – be it a lunar or planetary one. In the final analysis, detail must be regarded as consisting of just such contrasting effects and therefore any innovation(s) made in the optical system of the telescope which serves to enhance such contrasts will assist the detection of such detail.

Another fact which becomes evident to the observer resulting from this self-conditioning is that the dominant eye – the one normally and instinctively used – adjusts itself quickly and readily to the telescopic image, whereas the normally inactive member is noticeably inferior in performance as regards detection of detail and, moreover, finds the image bright, even painfully so, much as non-observers or novices do.

It would also appear that, regardless of general observing expertise, a person excels in his own specialised subject – the habitual Jupiter observer sees more on the planet than the lunarian and vice versa. Here again, brightness adaptation may play a significant role in this, plus greater familiarity with the *nature* of detail perceived.

In the early days of my observing I used a $6\frac{1}{2}''$ reflector to good effect but stepped up the aperture of my instrumentation as time went on. After using a $12''$ for many years and, latterly, a fine $10''$, I had occasion not so very long ago to use my old telescope, now belonging to a friend and, notwithstanding the excellent condition of the aluminised $6\frac{1}{2}''$ mirror, I felt positively handicapped by the apparent dimness of the image in the smaller instrument although using magnifications strictly proportionate to those in my present telescope. In fact, I saw nothing like the same amount of detail I had previously been able to record with the $6\frac{1}{2}''$ when it was in my possession. Advancing years may provide some explanation for this, but I have a feeling that, with continued *exclusive* use of the $6\frac{1}{2}''$, the brightness adaptation which would eventually take place might lead to the recovery, in part, of my previous success with this instrument . . . it would be an interesting experiment.

The lesson to be learnt from all these considerations is that no aspiring observer need feel outclassed by the larger, 'superior' equipment of his colleagues. It must not be forgotten that large telescopes are more prone to atmospheric vagaries besides being much less easy to operate – all things considered, it is the person 'at the small end' who really counts.

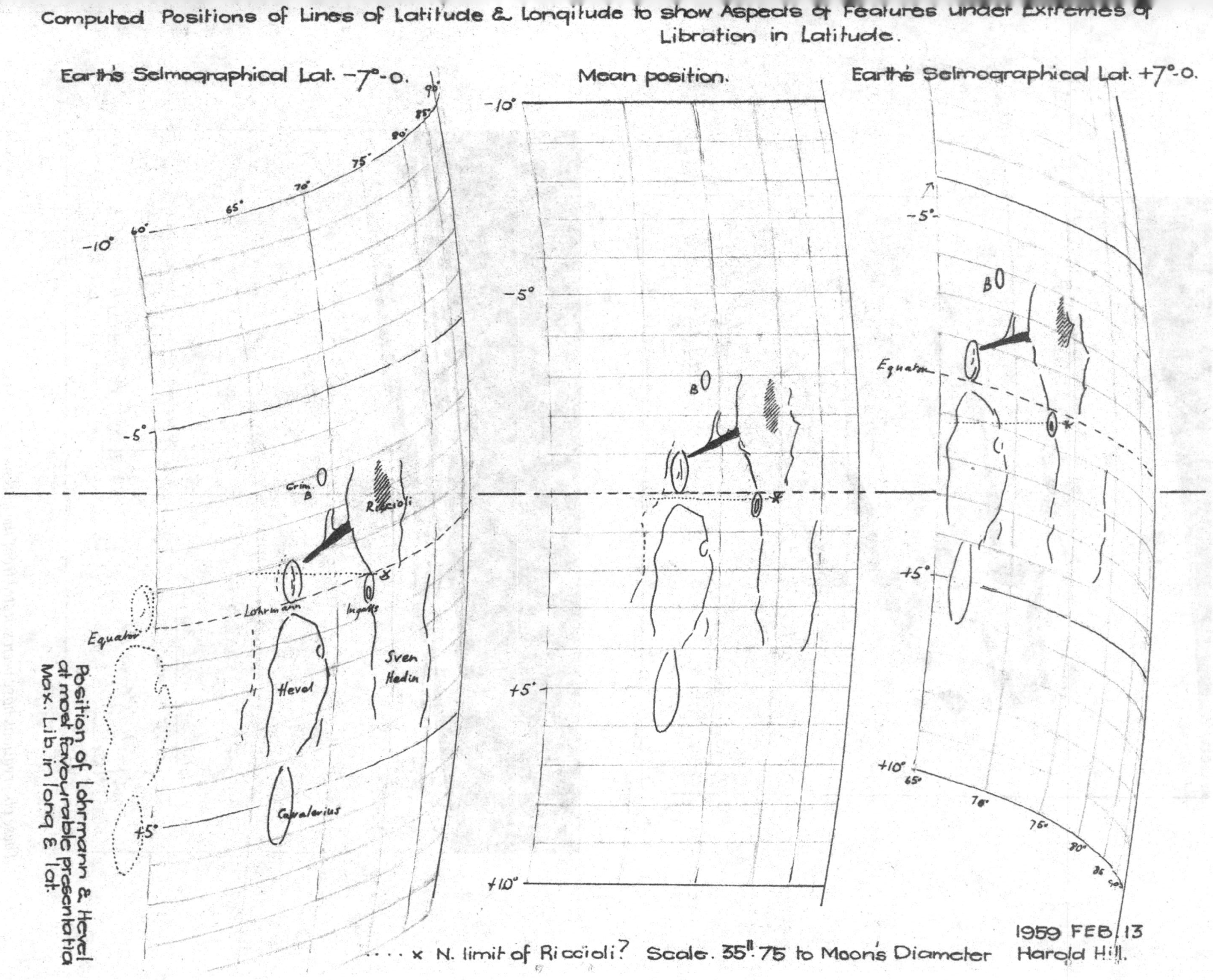

Computed Positions of Lines of Latitude & Longitude to show Aspects of Features under Extremes of Libration in Latitude.

Earth's Selenographical Lat. +7°·0.

Mean position.

Earth's Selenographical Lat. -7°·0.

B

Equator

-5°

+5°

+10°

65°

70°

75°

80°

Sco

1959 FEB. 13
Harold Hill.

-10°

-5°

+5°

+10°

N. limit of Riccioli?   Scale. 35″.75 to Moon's Diameter

Riccioli

Sven
Hedin

Crm.
B

Riccioli

Lohrmann

Hevel

Cavalerius

Equator

-10°

-5°

+5°

60°

65°

70°

75°

80°

85°

Position of Lohrmann & Hevel
at most favourable presentation
Max. Lib. in long & lat.

xxi

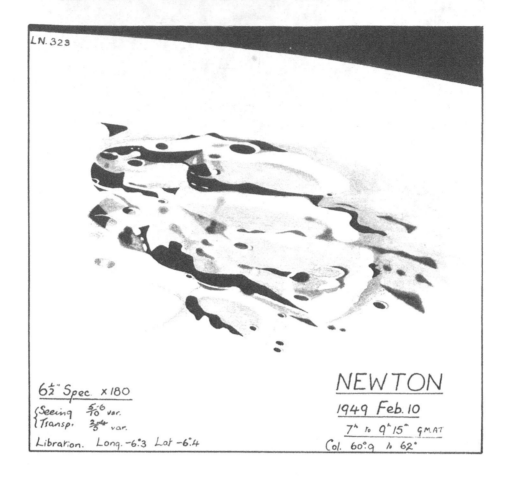

LN. 323

$6\frac{1}{2}''$ Spec. × 180

Seeing $\frac{5-6}{10}$ var.
Transp. $\frac{3-4}{5}$ var.

Libration. Long. −6°3 Lat −6°4

NEWTON

1949 Feb. 10

7^h to $9^h 15^m$ G.M.A.T.

Col. 60°9 to 62°

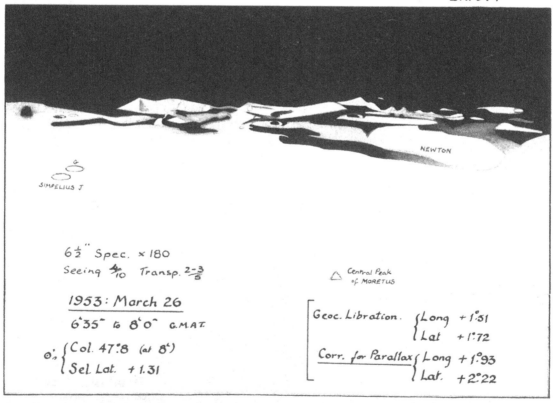

LN. 374

SIMPELIUS J

NEWTON

$6\frac{1}{2}''$ Spec. × 180

Seeing $\frac{4}{10}$ Transp. $\frac{2-3}{5}$

△ Central Peak of MORETUS

1953: March 26

$6^h 35^m$ to $8^h 0^m$ G.M.A.T.

θ_n { Col. 47°8 (at 8^h)
Sel. Lat. + 1°31

Geoc. Libration. { Long + 1°51
                   Lat + 1°72

Corr. for Parallax { Long + 1°93
                     Lat + 2°22

*These observations have been selected from the author's
early files to show the pictorial appearance of Newton
under such conditions.*

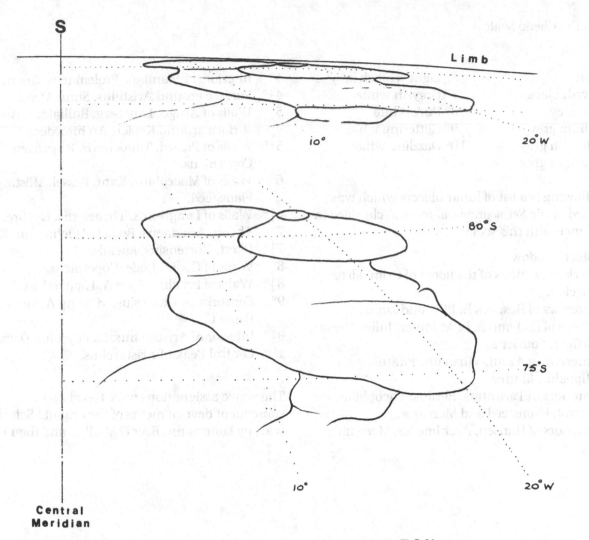

S

Limb

10°   20°W

80°S

75°S

10°   20°W

Central
Meridian

The complex formation NEWTON –
showing change of aspect at extremes
of northern & southern libration in latitude

The southernmost outline merely shows the degree of
foreshortening undergone when Newton is nearest to
the limb. No attempt has been made to show the vertical
relief displacement of this formation.

Schröter's Albedo Scale

0° Black
1° Greyish black
2° Dark grey
3° Medium grey
4° Yellowish grey
5° Pure light grey
6° Light whitish grey
7° Greyish white
8° Pure white
9° Glittering white
10° Dazzling white

The following is a list of lunar objects which was published in the *Selenographical Journal*, classified in accordance with this scale:

0°   Black shadows.
1°   Darkest portions of the floors of Grimaldi and Riccioli.
1½°   Interiors of Boscovich, Billy and Zupus.
2°   Floors of Endymion, Le Monnier, Julius Caesar, Crüger, Fourier a.
2½°   Interiors of Azout, Vitruvius, Pitatus, Hippalus, Marius.
3°   Interiors of Taruntius, Plinius, Theophilus, Parrot, Flamsteed and Mercator.
3½°   Interiors of Hansen, Archimedes, Mersenius.
4°   Interiors of Manilius, Ptolemaeus, Gueriké.
4½°   Surface around Aristillus, Sinus Medii.
5°   Walls of Arago, Lansberg, Bullialdus. Also surface around Kepler, Archimedes.
5½°   Walls of Picard, Timocharis. Rays from Copernicus.
6°   Walls of Macrobius, Kant, Bessel, Mösting, Flamsteed.
6½°   Walls of Langrenus, Theaetetus, La Hire.
7°   Theon, Ariadaeus, Bode B, Wichmann, Kepler.
7½°   Ukert, Hortensius, Euclides.
8°   Walls of Godin, Bode, Copernicus.
8½°   Walls of Proclus, Bode A, Hipparchus C.
9°   Censorinus, Dionysius, Mösting A, Mersenius B and C.
9½°   Interior of Aristarchus, La Péyrouse Delta.
10°   Central Peak of Aristarchus.

The above assignations were based upon subsequent developments of the original Schröter scale by Lohrmann, Beer & Mädler, and then Elger.

The charts

The introductory or key chart shows the arrangement of the 16 sectional maps on the four Quadrants which comprise the Earth-facing hemisphere.

The orientation shows south at the top as in the inverting astronomical telescope, with *east* to the left in conformity to the rules laid down by the Lunar Commission of the International Astronomical Union, instead of *west* used on all the old charts.

The sectional charts have been drawn on a scale of approximately 18 inches to the lunar diameter. In order to readily identify the location of each of the regions illustrated, these are indicated on the charts in heavy outline, together with their assigned regional number in the margins in order of appearance in the portfolio. To avoid the inconvenience of three-figure numbers in the circles, the numbers recommence with each of the four Quadrants.

In the interests of clarity I have avoided the insertion of the smaller details on the charts and, for the same reason, have confined myself to naming only a limited number of formations.

The columns indicated MT and ET with following figures denote the colongitudinal position of the morning and evening terminators, respectively. Other marginal numbers show the lines of latitude and longitude at 10 degree intervals for the mean librational position of the disc.

The scale of the drawings

In some instances where a possible reduction in the scale of a drawing from its original size might have to be made for reproduction purposes, a line with 5-mile intervals has been introduced in some cases. Where no such reduction was deemed to be necessary, the scale in terms of the lunar diameter has been given in inches.

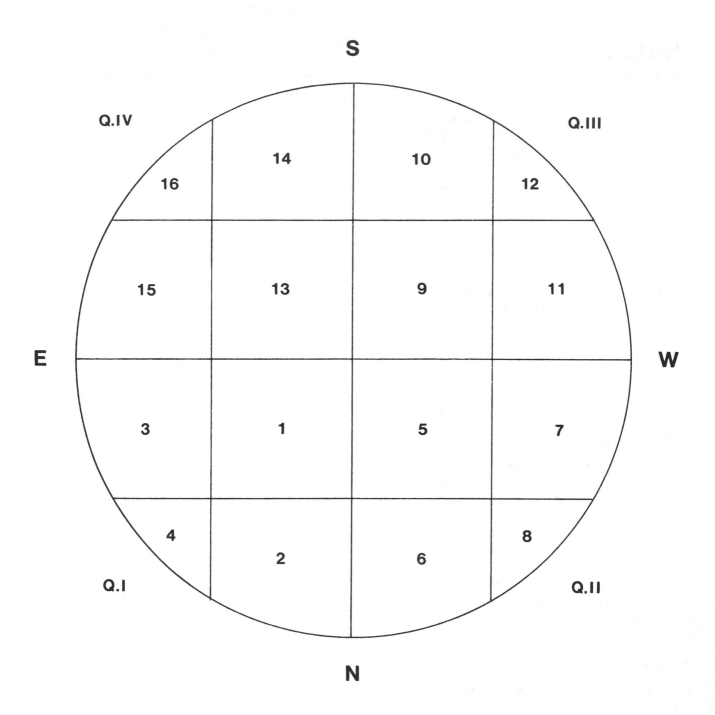

S

Q.IV

Q.III

14

10

16

12

15

13

9

11

E

W

3

1

5

7

4

8

2

6

Q.I

Q.II

N

2

Quadrant I – Section 1

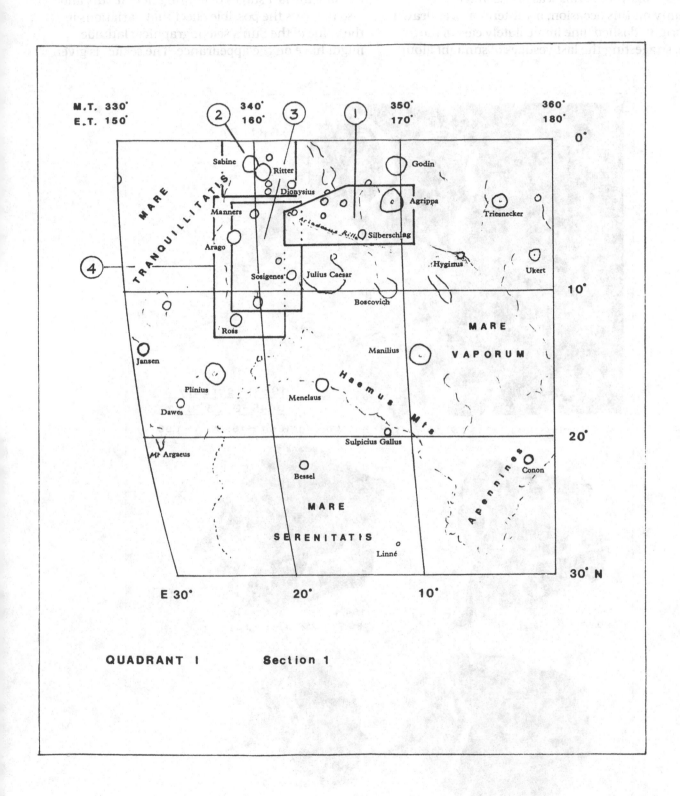

QUADRANT I   Section 1

3

The Ariadaeus Rille (western portion) at sunset

One of the objects of the 1983 Feb.4 observation was to ascertain if the Ariadaeus rille was traceable through the ridge which runs north from Silberschlag. Whilst this was not seen with certainty on this occasion, my attention was drawn to a bright 'dashed' line immediately east of the ridge, suggesting the last vestiges of sunlight along the northern border of the rille at this point. The four additional sketches (a) to (d) were made on later dates, and quite independently of each other, to study the last stages of lighting hereabouts and also to assess the possible effect that variations in the value of the Sun's selenographical latitude might have on the appearance. The sel.lat. is given

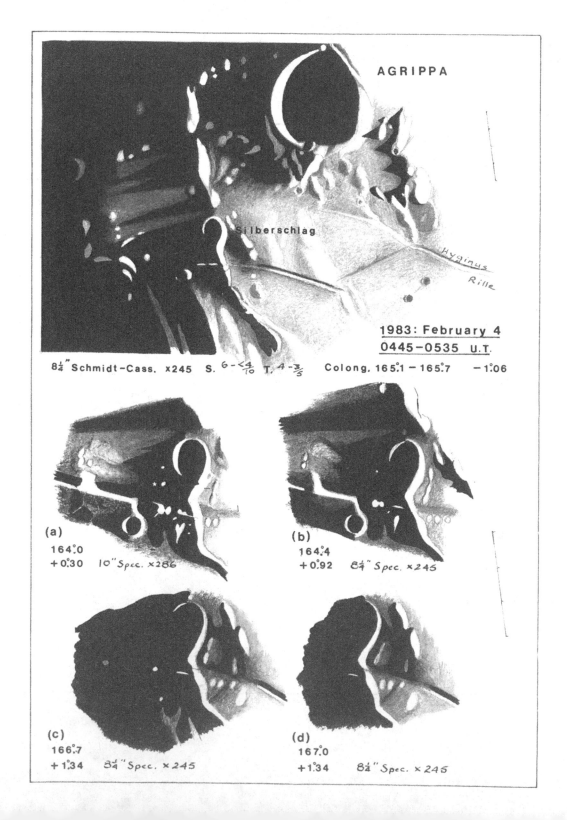

AGRIPPA

Silberschlag

Hyginus Rille

1983: February 4
0445–0535 U.T.

8¼" Schmidt-Cass. x245   S. 6–<4/10   T. 4–3/5   Colong. 165°.1 – 165°.7   –1°.06

(a)
164°.0
+0°.30   10" Spec. x286

(b)
164°.4
+0°.92   8¼" Spec. x245

(c)
166°.7
+1°.34   8¼" Spec. x245

(d)
167°.0
+1°.34   8¼" Spec. x245

beneath the colongitude figure for each of the sketches. No precise intercomparison of the drawings as regards colongitude is possible, unfortunately, but it is feasible that directional changes in the incident light caused by the slight lunar seasons may be responsible for some of the differences observed, notwithstanding the fact that this region is situated only 7° north of the lunar equator. A case in point is demonstrated by the two specks of light NE of Silberschlag in sketch (a) – the larger of which is marked by an asterisk for identification. Although the lighting is more advanced by some 50 min in sketch (b) the *smaller* of the pair in the south remains but the larger has gone whereas the reverse would be rather expected to occur. However, it may be significant that in (a) a notch was recorded in the crest of the Silberschlag ridge directly opposite the asterisked marking through which the last rays of the Sun may be striking, whereas in (b) under *different* circumstances of the Sun's sel.lat., the sunlight passes through the 'notch' (pass?) in another direction and thereby fails to illuminate the feature in question.

The lesson to be learned from these considerations is that apparent inconsistencies which occur in the comparison of drawings do not always necessarily redound to the discredit of the observer. Very often an explanation can be found such as the example expounded *always providing that drawings are executed with the utmost attention to detail.*

Further notes

In the localised sketches (c) and (d) it will be seen that the Ariadaeus rille was just traceable cutting through the ridge although the actual course had a partially filled-in appearance for a distance of some 8 miles west of the ridge, after which it became much more distinct on turning to run WNW. A dusky linear branch-rille was seen to run WSW to join the Hyginus rille. (see 1983 Feb.4 drawing)

The drawing below shows a short section of the eastern portion of the Ariadaeus rille at early sunrise with its southern banks brightly lit.

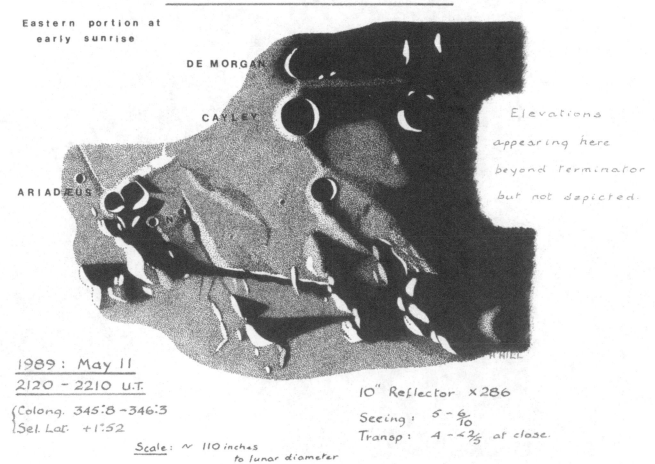

THE ARIADÆUS RILLE

Eastern portion at early sunrise

DE MORGAN

CAYLEY

ARIADÆUS

Elevations appearing here beyond terminator but not depicted.

A' RILLE

1989: May 11
2120 – 2210 U.T.
{ Colong. 345°8 – 346°3
{ Sel. Lat. +1°52

10" Reflector ×286
Seeing: 5 – 6/10
Transp: 4 – <2/5 at close.

Scale: ~ 110 inches to lunar diameter

The Sabine and Ritter Region at the evening terminator

The SW 'corner' of the Mare Tranquillitatis contains many features worthy of close study quite apart from the fact that it presents a fine spectacle when seen under morning and evening illumination.

The drawings opposite do not purport to show anything 'new' or of particular note but are included here to show the rapid lengthening of shadows during the observing session from the craters Manners and Arago B towards the marial ridges at the terminator and also to show the exaggerated prominence which these quite low elevations assume under very oblique lighting conditions. The penumbral effects from the crater shadows are evident where these are thrown athwart the ridges but, of course, the timing of such appearances is quite critical.

The rilles which run southwards towards Ritter from the Ariadaeus region are shown on this occasion to converge to a nodal point marked by a white spot (craterlet?) just west of Ritter C. At other times, usually under morning illumination, the eastern member of this pair of rilles has been seen to converge less sharply and pass the white spot/craterlet on its eastern side and other observers have seen it thus, so it would seem that the present account is in error. Some impediment to distinct vision must also have been operating on the same date because the diagonal rille running from Sabine towards the ruined formation Ariadaeus E was missed – probably due to indifferent seeing. For the same reason, the small object lying just NE of Arago B was twice depicted as a hill and yet has been observed as an unmistakeable craterlet at other times! Seeing can play some strange tricks with interpretation and I have found under a low lighting that the bright Sun-facing inner walls of craterlets can sometimes be mistaken for hills. In passing, it may be mentioned that, when near the terminator, craterlets with marked outer glacis can often appear quite disproportionately larger than when seen under a higher Sun.

Worthy of note is the considerable departure from circularity of the craters Schmidt, Ritter B and C and also Dionysius.

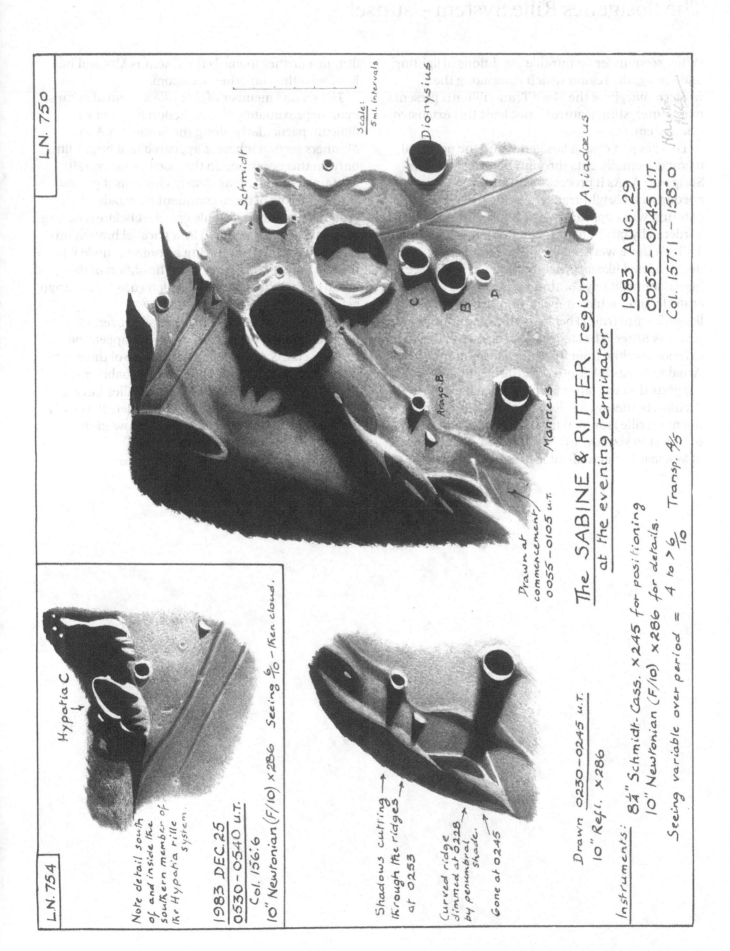

LN. 750

Scale: 5 ml. intervals

Schmidt

Dionysius

Ariadaeus

C

B

D

Arago B

Manners

Drawn at commencement 0055 - 0105 u.t.

The SABINE & RITTER region
at the evening Terminator

1983 AUG. 29
0055 - 0245 U.T.
Col. 157:1 - 158:0

LN. 754

Hypatia C

Note detail south of and inside the southern member of the Hypatia rille system.

1983 DEC.25
0530 - 0540 u.t.
Col. 156:6
10" Newtonian (F/10) x286 Seeing 6/10 - Thin cloud.

Shadows cutting through the ridges at 0253

Curved ridge dimmed at 0228 by penumbral shade.

Gone at 0245

Drawn 0230-0245 u.t.
10" Refl. x286

Instruments: 8¼" Schmidt-Cass. x245 for positioning
10" Newtonian (F/10) x286 for details.
Seeing variable over period = 4 to >6/10 Transp. 4/5

The Sosigenes Rille System – sunset

When seen under favourable conditions of lighting and seeing, the region which runs along the western margin of the Mare Tranquillitatis presents many interesting features – not least this extensive rille system.

This lies just east of Sosigenes and the principal member actually cuts through the smaller crater Sosigenes A. As it proceeds south from A it encounters a bright transverse line (rille?) before entering what appears to be a short cutting or defile bordered by bright banks. Later on (at colongitude 159°.4) these were seen to be beautifully disconnected like a string of elongated beads (see inset). Immediately south the rille crosses a distinct craterlet after which it meets another bright E–W linear feature (rille?) before continuing southwards towards Ritter. The stepped appearance of the exterior shadow from the ruined formation Ariadaeus E as it lies athwart the bright line suggests that the latter is an elevated bank – perhaps bordering a rille. The northward course of the main rille is shown in the present observation to be lost in shadow thrown from an elevation west of Maclear but actually it is traceable for some distance further towards the Haemus Mts and has been seen thus on other occasions.

The second member of this rille system also runs in an approximately N–S direction but is more difficult, particularly along the Sosigenes A to Manners section where it appeared as a bright line only on this occasion. To the north, however, its contained shadow was clearly visible as it passed through a row of three confluent ellipsoidal craterlets and then a single craterlet before entering Maclear. What appears to be a parallel branch on the western side and sharing a common node with the trio may, in reality, be a continuation of the broad coarse shallow valley which runs SE to Arago before it is lost at the terminator.

A footnote to the drawing refers to several features being recorded which do not appear on existing charts of the area. The nature of the bright linear streaks is doubtful but they probably mark the course of rilles. Obviously, those rilles having a W–E disposition are more difficult to detect as such because of the lack of detectable shadow in the direction of the incident light.

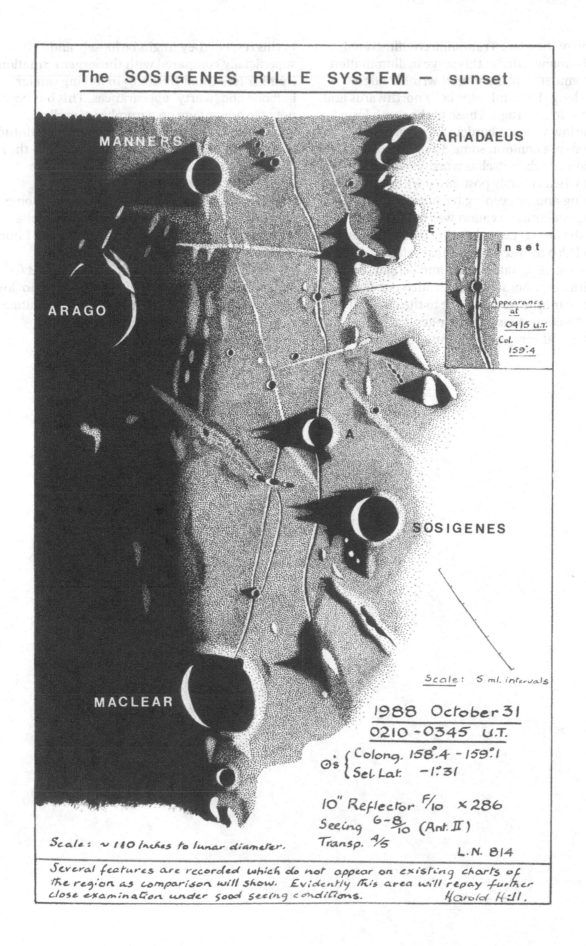

The SOSIGENES RILLE SYSTEM — sunset

MANNERS

ARIADAEUS

E

inset

ARAGO

Appearance
at
0415 u.t.
Col. 159°.4

A

SOSIGENES

Scale: 5 ml. intervals

1988 October 31

0210 - 0345 U.T.

⊙'s { Colong. 158°.4 - 159°.1
      Sel. Lat. - 1°.31

10" Reflector F/10 ×286

Seeing 6-8/10 (Ant. II)

Transp. 4/5

L.N. 814

MACLEAR

Scale: ~180 inches to lunar diameter.

Several features are recorded which do not appear on existing charts of the region as comparison will show. Evidently this area will repay further close examination under good seeing conditions.

Harold Hill.

The Arago to Ross Region

The region of the Mare Tranquillitatis illustrated opposite is noteworthy at this stage of illumination for the prominence of the marial wrinkle ridges running along the terminator both northwards and southwards from Arago. Those to the south form part of the low western border of the partially submerged ring Lamont, some 55 miles in diameter, a formation which in itself is worthy of individual attention when suitably positioned with regard to the morning and/or evening terminator.

This particular observation was made primarily to follow the last stages of lighting on the aforesaid ridges and also to observe the sunset effects on the Arago domes, designated alpha and beta in the present drawing. Actually, under such conditions, the latter bear little resemblance to the classical concept of a dome owing to their general asymmetry and complexity of shape and structure.

In this respect they might be loosely and superficially compared with the larger formation Rumker (see Quad.II Sect.8) in having similar bulbous and 'warty' appearances. This has been noted to advantage on a number of earlier occasions under a somewhat higher solar altitude but here the irregularities are only hinted – they demand better seeing conditions and larger aperture for effective study.

Minor swellings (domes?) are indicated some distance to the north and west of Arago alpha; these became much more conspicuous one hour later when at the terminator (see inset).

The effect of the speedy advance of sunset (or sunrise) is especially noticeable on features of low elevation such as wrinkle ridges and necessitates rapid working if the drawing is to be truly representative of the scene as a whole.

The ARAGO to ROSS Region at the Evening Terminator

LN. 793

MANNERS

β

ARAGO

α

MACLEAR

ROSS

Scale: 5 ml. intervals

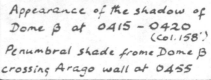

Appearance of the shadow of Dome β at 0415 – 0420 (Col. 158°.)
Penumbral shade from Dome β crossing Arago wall at 0455

1987 February 19
0210 – 0305 U T
Oʼs Col. 157°.0 – 157°.4

8¼" Schmidt-Cass. ×166 & ×245
Seeing good but variable 5 – 7/10
Transparency 4/5   (Ant. III - II)

Quadrant I – Section 2

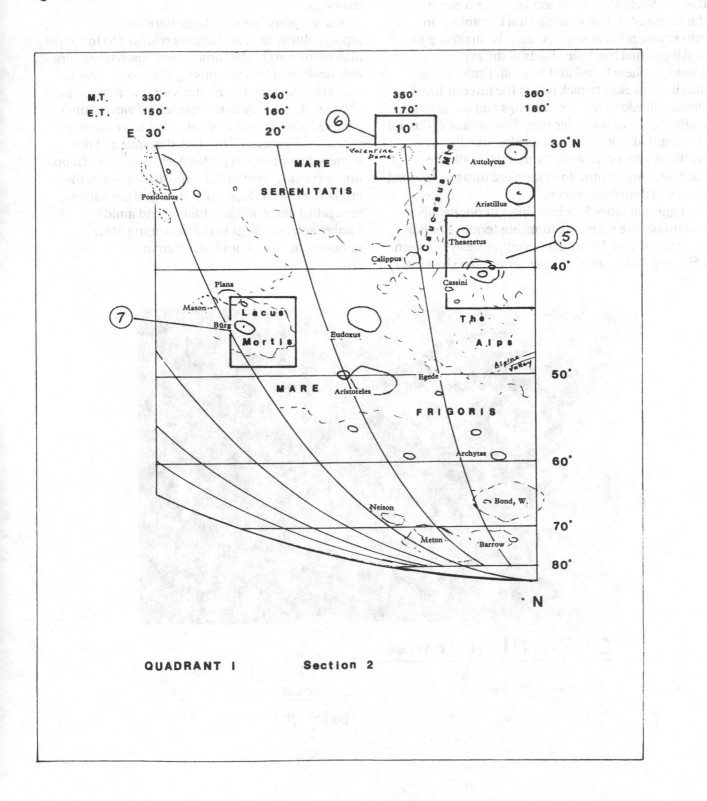

M.T.	330°	340°	350°	360°
E.T.	150°	160°	170°	180°

E 30° 20° 10° 30°N

⑥

"Valentine" Dome

MARE

SERENITATS

Posidonius

Caucasus Mts

Autolycus

Aristillus

Calippus

Theaetetus

⑤

Cassini

Plana

Mason

⑦

Lacus

Bürg

Mortis

Eudoxus

The

Alps

40°

Egede

Alpine Valley

50°

MARE

Aristoteles

FRIGORIS

60°

Archytas

Neison

Bond, W.

70°

Meton

Barrow

80°

N

QUADRANT I   Section 2

Cassini and surrounds

The two low-sun studies of this remarkable formation were made in the same lunation and are included here to illustrate the prominence which this relatively shallow object assumes when near the terminator. Considering that Cassini is thrown into strong relief at such times, it is surprising to find it omitted from the charts of the early selenographers Hevel and Riccioli. Under a high Sun it is not easy to pick out as the interior has a similar albedo to its surroundings and the bright walls appear as a slender ring. The outline of Cassini is polygonal, comprising several markedly rectilinear sections, and, though the walls are narrow, the surrounding glacis are unusually broad and even double in places.

Suggested times for observing interior details to advantage are when the morning terminator lies between 5° and 10° and for evening study between 160° and 166°. During these periods the shadows are not intrusive, in contrast to the interior craters A and B which, like Theaetetus to the south, hold considerable shadow at such times – emphasising their depth.

It is not proposed to enlarge here on topographical facts and figures relevant to this most interesting part of the lunar landscape as these are ably dealt with in descriptive guides to the Moon, but rather to draw the reader's attention to a region of compelling attraction, especially if viewed under oblique lighting, and to show what is within reach of modest apertures. The stark desolation of the scenery hereabouts is particularly well brought out under evening light as the contrasts between the brilliant western slopes of craters and mountains, set against encroaching shadow and amid the sombre surroundings of the darkening Mare Imbrium, are harsh and spectacular.

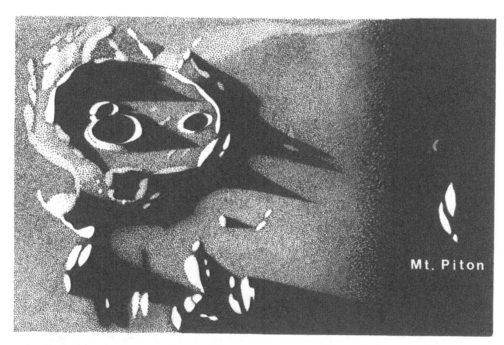

Mt. Piton

CASSINI at sunrise

8¼" Schmidt-Cass. ×245

Seeing: 4–6/10   Transp: 3/5

Geoc. Libr. (2⊙) L = +3°40 B = −5°24

1989 : January 14

2005 – 2035 U.T.

⊙'s { Colong. 359°9 – 000°3
      { Sel. Lat. −1°.03

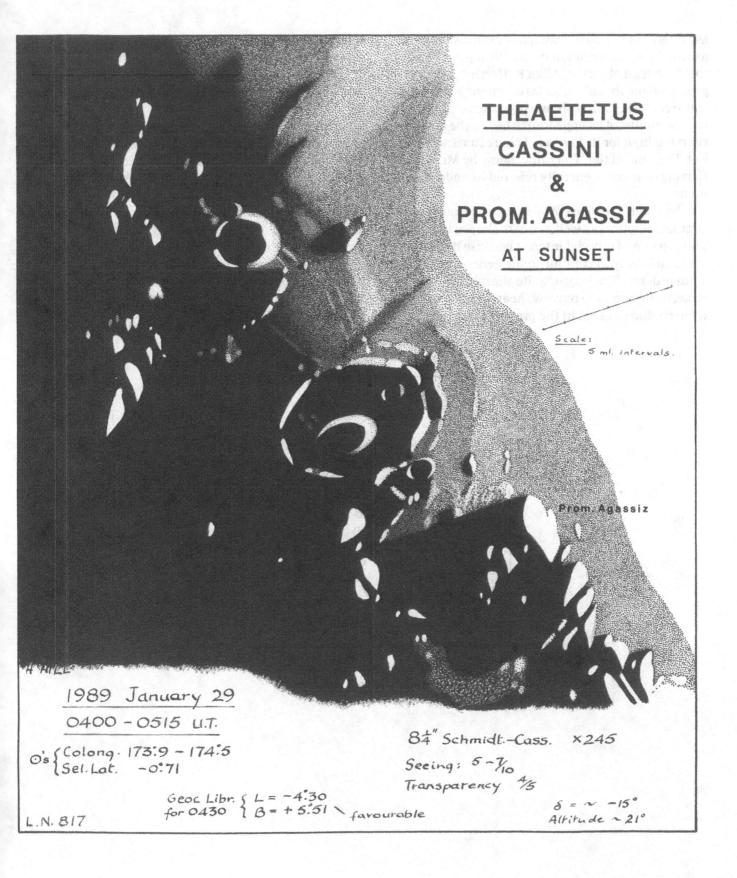

THEAETETUS
CASSINI
&
PROM. AGASSIZ

AT SUNSET

Scale:
5 ml. intervals.

Prom. Agassiz

H HILL

1989 January 29
0400 - 0515 U.T.

O's { Colong. 173°.9 - 174°.5
{ Sel. Lat. -0°.71

8¼" Schmidt.-Cass. ×245

Seeing: 5 - 7/10
Transparency 4/5

Geoc Libr. { L = -4°.30
for 0430 { B = + 5°.51 \ favourable

δ = ~ -15°
Altitude ~ 21°

L.N. 817

The Plateau-Dome NE of Linné

My attention was first drawn to this unusual feature by an observation dated 1966 June 25 by the American observer, Alika K. Herring, and this prompted me to make a series of drawings of this feature, of which that shown opposite is representative of its appearance before the Sun has risen too high for its low-relief characteristics to be lost. Designated the 'Valentine' dome by Mr Herring, it is now generally referred to under that name.

It has been described by observers variously as circular, elliptical as well as heart-shaped in outline. The object is of especial interest because it does not conform closely in contour to the generic notion of a lunar dome. The bright, quite sharply defined eastern edge and the narrow, heavily shaded western slope as seen in the present drawing suggests a relatively steep rise from the Mare surface before flattening off, although the shading of the plateau itself denotes a gentle convexity. A similar but reversed effect is seen under opposite lighting.

Mr Herring, using a 61″ reflector with a power of × 900 under 7 seeing saw a rille traversing the dome from SE–NW, together with several tiny craterlets on and around the plateau. Also he saw two short very delicate parallel rilles at the eastern end of the principal rille. Such details can hardly be expected to be within range of only moderate apertures but they do present a challenge to sharp-eyed observers with adequate instrumentation.

Herring's observation is reproduced here and was taken from the B.A.A.'s Lunar Section publication *The Moon* Vol.15 No.2.

Observation by Alika K. Herring
1966 June 25   61″ rfl. × 900

THE PLATEAU / DOME
N.E. OF LINNÉ [L.N.803]

S
E
W
(I.A.U)

10" Reflector F/10 ×190
×286
Seeing. 6-7/10 Transp. 2-3/5
Moon's Decl. -12°45'

1987 November 27
1720 - 1745 u.t.
⊙'s ⎰Col. 352°.7 - 352°.9
      ⎱Sel. Lat -1°.39

Lacus Mortis and Burg

This observation was made to check the extraordinary shadow profiles thrown by the western border of the Lacus Mortis – an effect first noticed by the author on 1985 November 3 at a very similar stage of illumination.

The general shapes of the shadow spires and their deviation in direction were fully confirmed proving that floor irregularities were responsible for these most unusual effects. Some of the floor shadings are puzzling in that they do not readily conform to topographical features seen under a higher evening Sun. Heavy shadow indicates pronounced slumping of the Lacus surface at the western foot of the ridge running NW from Burg and indeed to the south along the western glacis of Burg itself. At the time of observing it did not seem likely that this was due to the normal curvature of the Lacus surface because the edge of this shadow was fairly sharply defined.

The shadow spires are shown as drawn between 0130 and 0145, and from then onwards the shadows lengthened rapidly to merge with the darkening floor.

It is highly desirable that these appearances be further studied under a favourable combination of libration, good seeing and, importantly, air clarity. Much could be learned about surface inequalities from early morning aspects also; altogether a fascinating region with prominent fault lines south of Burg and a distinct rille which breaks through the western border of the Lacus and crosses the floor in a NE direction. These features are still illuminated in the drawing but are seen to better advantage under a higher Sun.

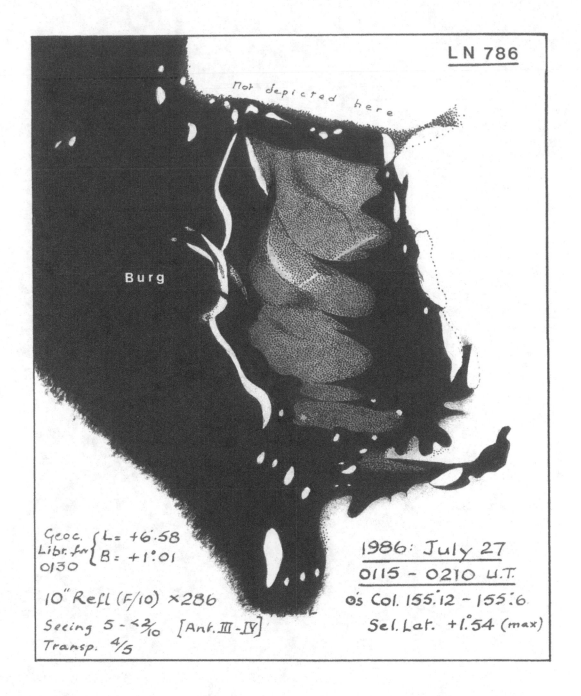

LN 786

not depicted here

Burg

Geoc.
Libr. for { L = +6°.58
0130   B = +1°.01

10" Refl (F/10) ×286

Seeing 5 - <²/₁₀ [Ant. Ⅲ-Ⅳ]
Transp. ⁴/₅

1986: July 27
0115 - 0210 U.T.

O's Col. 155°.12 - 155°.6

Sel. Lat. +1°.54 (max)

Quadrant I – Section 3

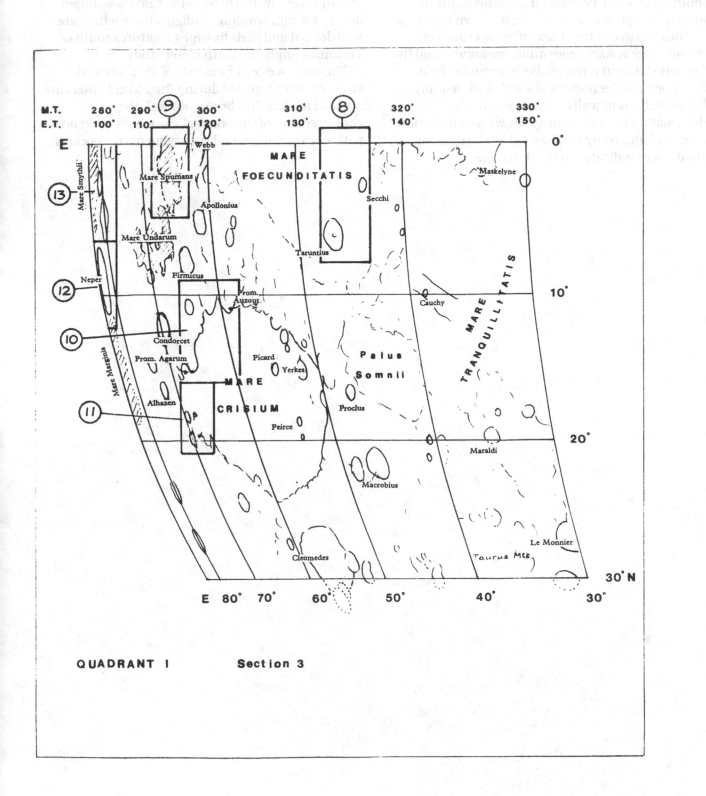

QUADRANT I   Section 3

The evening terminator from Messier A to Taruntius

Though not on my programme, the appearance of the topographical features in the neighbourhood of Secchi and the larger formation Taruntius on the morning of 1987 February 17 prompted me to attempt a representation which is shown opposite.

The diversity in the shape of lunar craters of medium size is sometimes quite remarkable and the formation Secchi is noticeably of irregular form, being open to the north and south with sensibly linear sections of wall on the western side. This deformation from circularity shows well under low evening light, being emphasised by the growing shadows as indicated in the drawing.

The marial ridges form an interesting subject although the advance of sunset at this stage of lighting demands rapid attention if they are to be shown correctly. It will be noted that these ridges display smooth winding configurations which are well defined and their linkage to features south of Taruntius might be worth closer study.

The outer western ramparts of Taruntius and environs were depicted during the second observing session; the low Sun brings out well the characteristics of the broad glacis which are scored with narrow valleys and enlargements or 'pockets'.

The Evening Terminator from

MESSIER A to TARUNTIUS

LN. 793

Messier A

Appearance of Messier A at 0405
Remains of Messier had gone.

Scale:
5 ml. intervals

Secchi

Taruntius

1987 February 17
0240 - 0300 UT
then
0310 - 0405 UT
Col. 133°.0 - 133°.7

Geoc. Libr.   { L = -5°.67
for   { B = -0.81
0330
               rather
               unfavourable

Scale: 90 inches to Lunar Dia.
       1 inch = 24 miles

{ Seeing 6-7/10 (Ant. II)
{ Transp. 4/5

8¼" Schmidt-Cass. ×166 & 245

Harold Hill

23

The Mare Spumans under advancing evening light

The Mare Spumans or 'Foaming Sea' is a relatively little known region and one that is rarely, if ever, the subject of representation by the amateur lunarian. This neglect is almost certainly due to the absence of any really noteworthy objects to portray and yet, as the Sun declines, the scene becomes one of absorbing interest as hitherto unsuspected features begin to assume ever-increasing prominence with dramatic spires of shadow developing to creep across the darkening surface of the Mare.

The crater which is positioned centrally on the eastern border of the Mare is Dubiago P and those in the foreground are identifiable as subsidiaries of the Apollonius and Maclaurin groups.

Mare Spumans is almost completely contained within the drawing and measures some 95 miles from N–S and about 40 miles from E–W. It crosses the lunar equator at a mean longitude of 65° E. In consequence, its presentation is affected chiefly by the degree of libration in longitude. In the present instance, conditions were rather unfavourable with some inevitable foreshortening.

Another of the lesser seas which might repay attention under low lighting and which also seems to have been scarcely observed in the past is the companion which has the unimaginative title of Mare Undarum ('Sea of Waves'!). This adjoins Mare Spumans on the NE and is of greater extent. There has been some speculation that both maria may be, in reality, groups of coalesced crater rings which at some epoch were flooded with basaltic material. Certainly the general outline of M. Undarum – especially when viewed under a high Sun – is highly suggestive of this, but it is less obvious in the case of M. Spumans which may have suffered a general flooding from its immediate and larger neighbour the Mare Foecunditatis which lies to the west.

L.N. 753

MARE SPUMANS
under advancing evening light

1983 NOV. 22
0130 - 0225 U.T.
Col. 113°.40 - 113°.85

Geoc. Libr. $\begin{cases} l = -4°.00 \\ b = -0°.28 \end{cases}$

10" Newtonian Refl. × 286   Seeing 6 - 7/10
                                Transp. 4/5

A. HILL

The South-Eastern 'Corner' of the Mare Crisium Region

Many pages of this portfolio could be devoted to discussion and illustration of the features of this compellingly attractive lunar sea and its rugged surrounds but, for reasons of space limitation, there are restrictions on what can appear.

Five sunset observations have been selected and arranged in order of advancing colongitude and presented to show the fascinating changes and, in some instances, puzzling effects which occur along the SE and S border from the Prom.Agarum to Prom.Auzout delta.

Some of the inconsistencies which are evident in the shapes and positions of the peaks receiving the last rays of the Sun may be due in part to variations in the direction of incident light coupled with the changing values of colongitude and, for this reason, the relevant data are added to each drawing. Geocentric librational values are also given for the mean period of each observing session – these cover a wide range of circumstances and materially affect the presentation of the areas shown in the five sketches.

A STUDY OF THE SUNSET SEQUENCE ON THE

PROM. AGARUM &

SE CORNER OF THE

MARE CRISIUM REGION

UNDER VARYING PRESENTATIONS
OF LIBRATION IN LAT. & LONG.
AND DIFFERENT SOLAR
SELENOGRAPHICAL LATITUDES

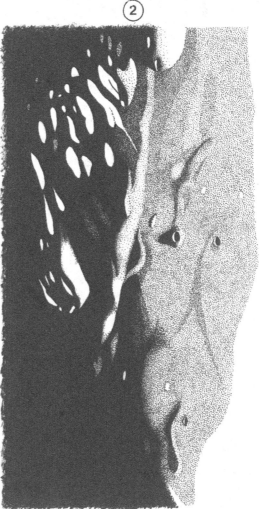

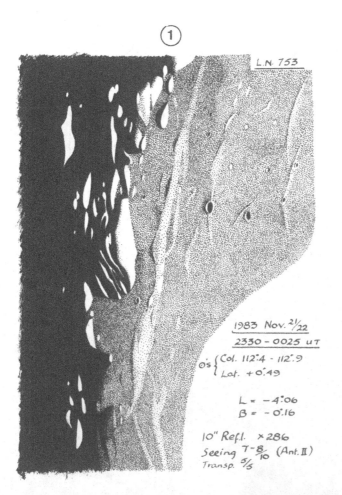

L.N. 753

1983 Nov. 21/22

2330 – 0025 UT

⊙'s { Col. 112°.4 – 112°.9
     { Lat. +0°.49

L = –4°.06
B = –0°.16

10" Refl. ×286
Seeing 7-8/10 (Ant. II)
Transp. 5/5

L.N. 756

1984 Feb. 19

0230 – 0330 UT

⊙'s { Col. 116°.1 – 116°.7
     { Lat. –1°.48

L = +3°.43
B = –6°.21

10" Refl. ×286
Seeing <6/10 then deteriorating.
Transp. 4/5

26

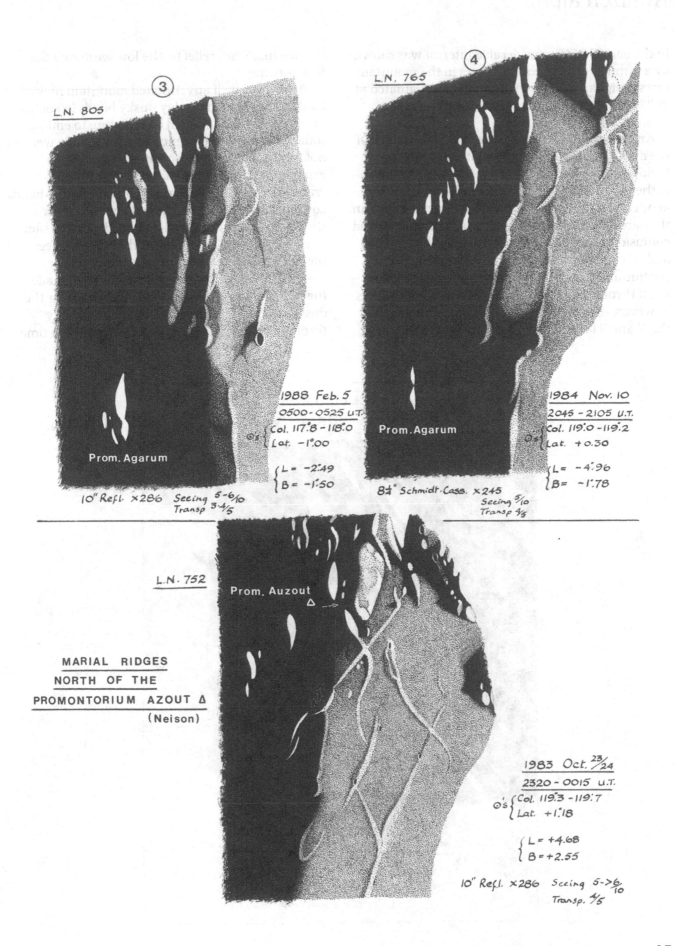

③

L.N. 805

1988 Feb. 5
0500 - 0525 U.T.
⊙'s { Col. 117°.8 - 118°.0
      { Lat. -1°.00

      { L = -2°.49
      { B = -1°.50

Prom. Agarum

10" Refl. ×286   Seeing 5-6/10
                  Transp 3·4/5

④

L.N. 765

1984   Nov. 10
2045 - 2105 U.T.
⊙'s { Col. 119°.0 - 119°.2
      { Lat. +0°.30

      { L = -4°.96
      { B = -1°.78

Prom. Agarum

8¼" Schmidt·Cass. ×245
              Seeing 5/10
              Transp 4/5

L.N. 752

Prom. Auzout
△ →

MARIAL RIDGES
NORTH OF THE
PROMONTORIUM AZOUT △
(Neison)

1983 Oct. 23/24
2320 - 0015 U.T.
⊙'s { Col. 119°.3 - 119°.7
      { Lat. +1°.18

      { L = +4°.68
      { B = +2°.55

10" Refl. ×286   Seeing 5→6/10
                 Transp. 4/5

Alhazen alpha

In the early 1970s considerable interest was shown by a number of amateur observers in the mountain mass designated Alhazen alpha which is situated at 20° N 70° E on the eastern margin of the Mare Crisium.

An attempt was made to determine the nature of the dusky bands radiating from its summit; the bands become visible soon after the western slopes of the mountain are clear of deep shadow, i.e. from about colongitude 333° onwards. Under a high Sun they are difficult to locate due to lack of shadow and confusion of detail in the general neighbourhood and it is only with the familiarity resulting from continuous observation that it is possible to follow them throughout a lunation. As sunset approaches, however, they become easy objects, particularly at the N and S faces of the mountain where they are

thrown into stark relief by the low westering Sun (see drawings).

There are few, if any, isolated mountain masses on the Moon which display dusky bands in such a marked form and it would be tempting to embrace some form of erosive agency as accountable were it not for the difficulties associated with such an explanation. It has been claimed that summit craterlets exist on Alhazen alpha and that the bands could be lava flows from these. However, there is strong evidence that they lie along depressed sites – see the shadow 'channels' at the N and S faces and the intrusion of shadow into the bands on the sunset drawings. The suggestion has been made that boulder-strewn valleys could account for the duskiness but this is negated by their visibility throughout the lunar day, especially at those times

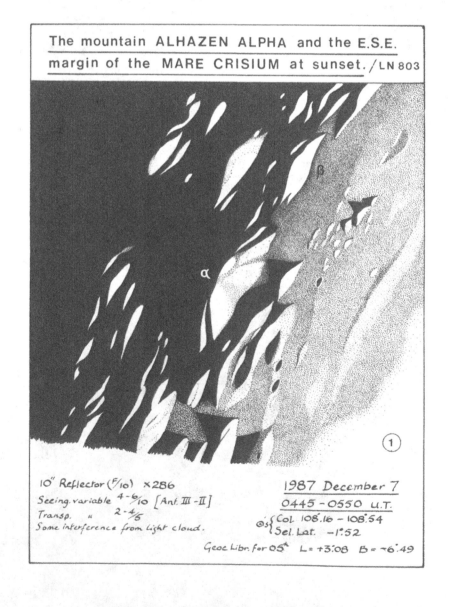

The mountain ALHAZEN ALPHA and the E.S.E. margin of the MARE CRISIUM at sunset. /LN 803

β

α

①

10" Reflector (F/10) ×286
Seeing variable 4-9/10 [Ant. III-II]
Transp.   „   2-4/5
Some interference from light cloud.

1987 December 7
0445 - 0550 U.T.
⊙'s { Col. 108°.16 - 108°.54
      { Sel. Lat. -1°.52
Geoc. Libr. for 05ʰ L = +3°.08 B = -6°.49

when contained shadow would disappear.

The dominating altitude of Alhazen alpha is effectively demonstrated in drawings (1) and (2) by its enormous evening shadow which cuts through and obliterates lesser elevations to the east – its presence is clearly outlined. The sites of the bands on the N and S slopes of the mountain are shadow-filled and the shadow outlines from elevations immediately west are encroaching on the lower slopes into the banded zones indicating that these are depressed sites. Subsequent stages are to be seen in drawing (2) at colongitude 109° where the effects are even more pronounced.

Note the darkening margins of the Mare at the base of mountains beta and gamma, the latter a designation given by the author on his own initiative, purely for the purpose of reference, since there is none on available maps.

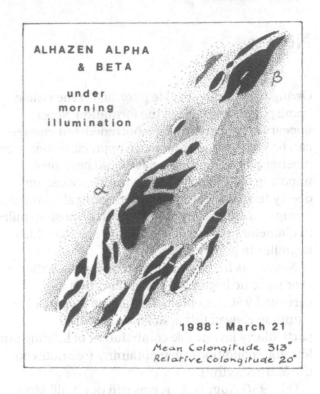

ALHAZEN ALPHA
& BETA

under
morning
illumination

1988 : March 21

Mean Colongitude 313°
Relative Colongitude 20°

The mountain ALHAZEN ALPHA and the E.S.E. margin of the MARE CRISIUM at sunset. / LN 803

10" Reflector (F/10) ×286
Seeing variable 5-7/10
Transp. 4/5

1987 December 7
0625 - 0700 U.T.
⊙'s { Col. 109°.0 - 109°.3
      { Sel Lat. -1°.52
Geoc. Libr. for 07ʰ L = +3.00 B= -6°.46

Neper

Owing to its unfavourable position on the visible hemisphere (84° E long) the formation Neper appears at best greatly foreshortened, but enough can be seen for the observer to appreciate that, were it better placed on the disc, it would be a most imposing object with its lofty, multi-peaked and deeply terraced walls, prominent central mountain group and sizeable interior craters. Almost 90 miles in diameter, it is larger than Langrenus to which it is similar in generic type.

Neper has had a special interest for the writer ever since he began serious observation and, as early as 1950, he obtained what was regarded as a unique view of this area because of the particularly favourable combination of lighting and libratory conditions then obtaining (see notes on the Mare Smythii).

On 1950 August 27 it was still not 'Full' Moon at the time of observation but, even at colongitude 88°.4, there was already a marked defect of illumination at the eastern side of the disc because of a strong E libration of more than 5°. In consequence, the terminator was visible over regions on the normally averted hemisphere. For instance, Jansky, positioned east of Neper at a mean longitude of 89° E was brought well into view and features still further east could be discerned. Under such favourable conditions, the central longitudinal ridge on the interior and the floor, marked by areas of different tone, were well displayed.

A repeat observation under almost precisely the same conditions was made 2 Saros cycles or 445 lunations later (!) but on this occasion the lighting allowed greater penetration into the 'hidden' hemisphere, as will be evident if the two drawings are compared. Inevitably, minor differences over some detail will be found, e.g. in the region SE of Neper, but in general, the similarity is close and reinforces what has been written elsewhere about the precise repetition of all the parameters after an 18-year interval.

The Neper region was surveyed at the intermediate 1968 epoch but cloud and/or bad seeing prevented effective observation on the relevant dates.

The region is, conversely, at its worst possible presentation at the intervening 9-year epoch with Neper's E wall actually beyond the limb with the interior completely hidden by the crater's W wall – the latter, in itself, a difficult object to identify!

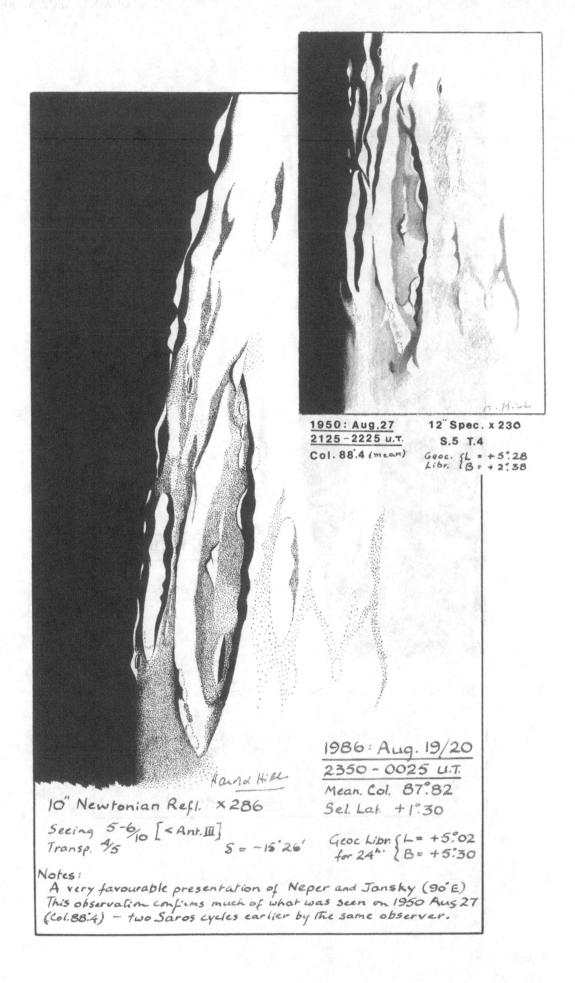

1950: Aug.27   12" Spec. x 230
2125 - 2225 U.T.   S.5 T.4
Col. 88.4 (mean)   Geoc. {L = + 5°.28
                      Libr. {B = + 2°.38

1986: Aug. 19/20
2350 - 0025 U.T.
Mean. Col. 87°.82
Sel. Lat. +1°.30

10" Newtonian Refl. × 286

Seeing 5-6/10 [< Ant. III]
Transp. 4/5   δ = -15°26'   Geoc Libr. {L = +5°02
                                         for 24ʰ. {B = +5°30

Notes:
  A very favourable presentation of Neper and Jansky (90°E)
This observation confirms much of what was seen on 1950 Aug 27
(Col. 88°4) - two Saros cycles earlier by the same observer.

LN 788

S

E — W

(I. A. U)

Scale:
5 ml. intervals

NEPER

– at sunset

1986 Sept. 18

2240 – 2305 U.T.

Col. 93°.36 – 93°.57

A HILL

8¼" Schmidt-Cass. ×245

10" Newtonian Refl. ×286

Seeing 7-8/10 [Ant. II]

Transp. 5/5

Geoc Libr { L = +5°.39 (max)
fv 23h   { B = +1°.92

   Outlines drawn with 8¼" S.C. but as seeing
improved 10" Newtonian used for the details.

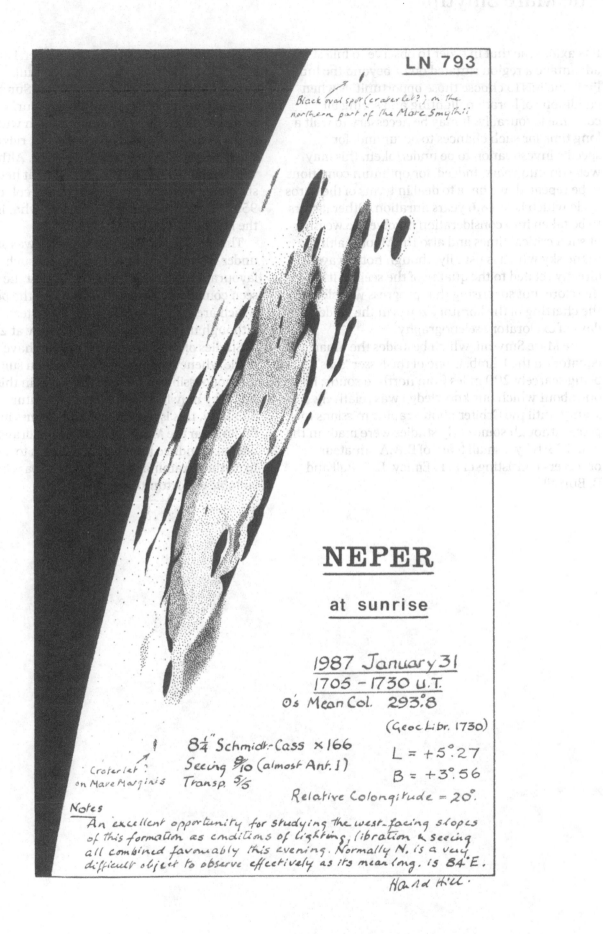

LN 793

Black oval spot (craterlet?) on the
northern part of the Mare Smythii

NEPER

at sunrise

1987 January 31
1705 - 1730 U.T.
O's Mean Col. 293°.8

(Geoc.Libr. 1730)

8¼" Schmidt-Cass ×166
Seeing 9/10 (almost Ant. I)
Transp. 5/5

Craterlet?
on Mare Marginis

L = +5°.27
B = +3°.56

Relative Colongitude = 20°.

Notes
An excellent opportunity for studying the west-facing slopes
of this formation as conditions of lighting, libration & seeing
all combined favourably this evening. Normally N. is a very
difficult object to observe effectively as its mean long. is 84°E.

Harold Hill.

The Mare Smythii

It is axiomatic that in order to observe to fullest advantage a region situated on or beyond the lunar limb one has to choose those opportunities when conditions of libration, lighting and seeing all combine favourably. It may be necessary to wait a long time for such chances to occur and, for a specific investigation to be undertaken, this may well run into years. Indeed, for optimum conditions to be repeated, we have to deal in terms of the Saros cycle which is of 18.6 years duration. Other factors to be taken into consideration involve the weather at such critical times and also the Moon's altitude in the sky which is usually, though not always, directly related to the quality of the seeing. It is, therefore, not surprising that progress was slow in the charting of the libratory zones in the golden days of exploratory selenography.

The Mare Smythii, which bestrides the lunar equator on the E limb, is one of the lesser 'seas' being scarcely 200 miles from north to south; it is one about which our knowledge was relatively scanty until the Orbiter photographic missions took place although some early studies were made in the mid-1930s by a small team of B.A.A. amateur observers consisting of E. F. Emley, L. F. Ball and B. Burrell.

The observation of 1986 September 18 which is illustrated here was made just after Full Moon with strong libration from the E limb. The Sun has set on the eastern side of the Mare and the surface which remains visible shows that it is strewn with craters, craterlets, a number of obscure rings, ridges and small hills – all greatly foreshortened. Although the western border seems to be of no great height, the shadows encroach rapidly so that by colongitude 95°–96° they are pronounced, and what is left of the Mare is very dark.

The drawing of 1987 January 31 was secured under limiting conditions but under such favourable libration (L = + 5°.24) that the Mare is seen completely within the limb and the peaks which are seen in profile mark its eastern border although they no longer hold shadow at 294°.

Studies of these mountain profiles have been made whenever conditions have been suitable but it is not possible to include the series in this portfolio and they may form the subject of a future treatise. Similarly, this is not the place to discuss in detail the findings on the Mare surface but the drawings are offered to induce interested observers to devote time to further examination of Smyth's Sea whenever suitable opportunities occur.

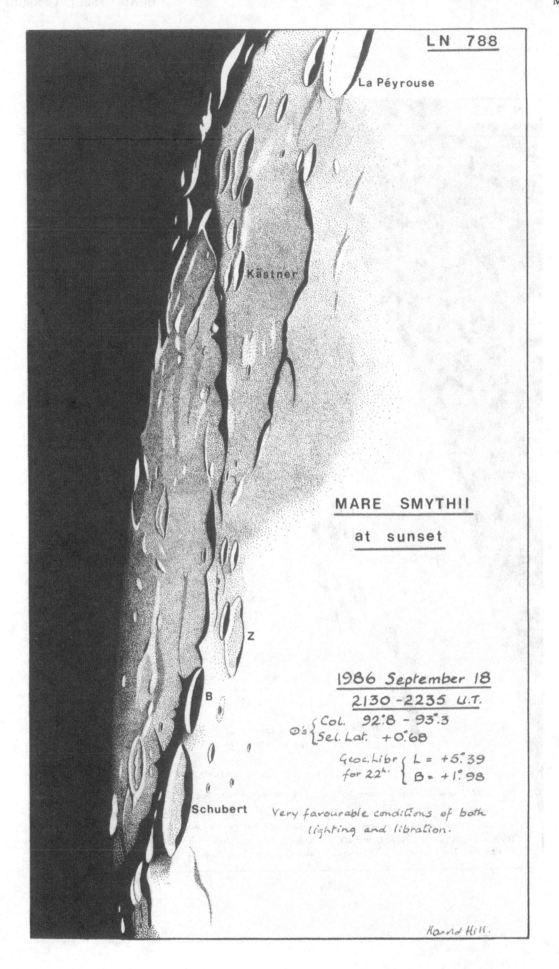

LN 788

La Péyrouse

Kästner

MARE SMYTHII

at sunset

Z

B

1986 September 18

2130 – 2235 U.T.

⊙'s { Col. 92°.8 – 93°.3
{ Sel. Lat. +0°.68

Geoc. Libr { L = +5°.39
for 22ʰ. { B = +1°.98

Schubert   Very favourable conditions of both
lighting and libration.

Harold Hill.

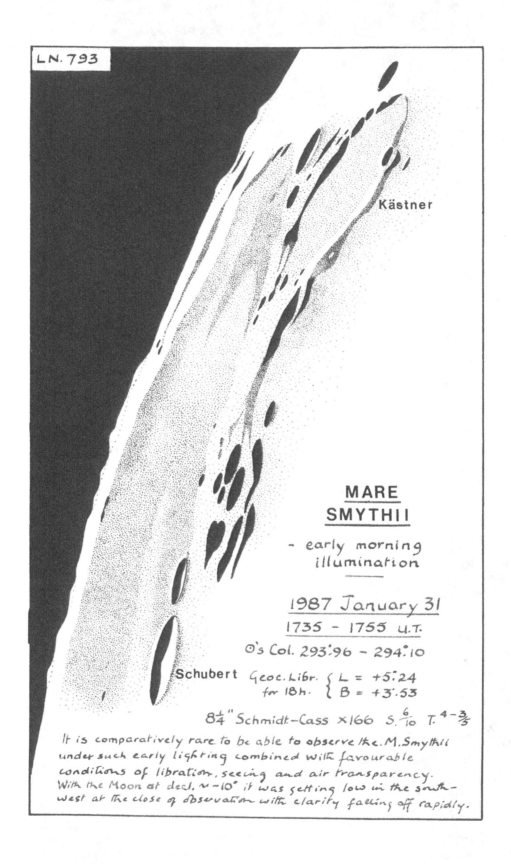

LN. 793

Kästner

MARE
SMYTHII

- early morning
illumination

1987 January 31
1735 - 1755 U.T.

⊙'s Col. 293°96 - 294°10

Schubert   Geoc. Libr. { L = +5°24
for 18h.   { B = +3°53

8¼" Schmidt-Cass ×166 S. $\frac{6}{10}$ T. $4 - \frac{3}{5}$

It is comparatively rare to be able to observe the M. Smythii
under such early lighting combined with favourable
conditions of libration, seeing and air transparency.
With the Moon at decl. ~ -10° it was getting low in the south-
west at the close of observation with clarity falling off rapidly.

Quadrant I – Section 4

Quadrant I – Section 4

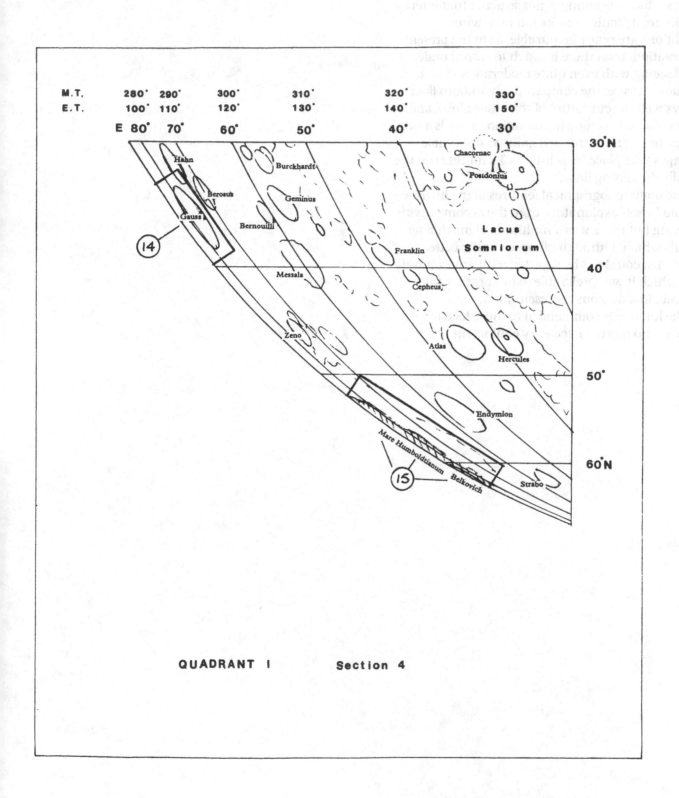

Gauss. The walled-plain at sunset

A magnificent walled-plain, fully 110 miles in diameter at mean position 36° N 79° E which would be a most imposing object if more centrally placed. Even so, foreshortening is not so acute that interior details are difficult to make out and, when conditions are really favourable, as in the present observation, then there is much to record under good seeing with even quite moderate aperture.

Due to its size, the comparatively smooth floor shows well the curvature of the lunar globe, and, under oblique evening illumination, there is a need for haste when recording details, as noticeable changes take place in as little as 15 minutes on the rapidly darkening floor.

The main topographical features of the interior should be self-explanatory from the accompanying drawing but there was a multitude of much finer details which, although glimpsed in the better moments, could not be depicted with any certainty and which it was preferable to omit rather than indicate in a dubious representation.

The interior is commended to the interested reader who has the necessary equipment.

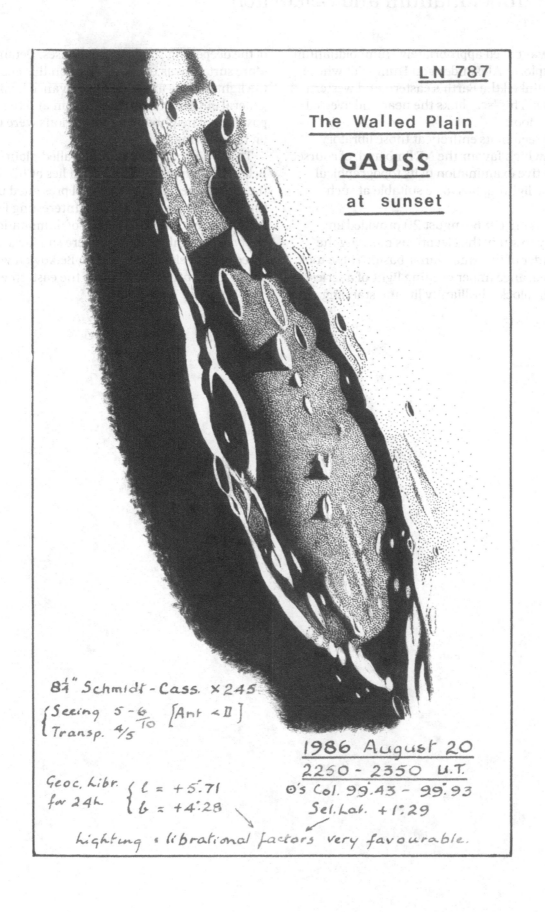

LN 787

The Walled Plain

GAUSS

at sunset

8¼" Schmidt - Cass. ×245

{ Seeing 5-6/10 [Ant < II]
{ Transp. 4/5

1986 August 20

2250 - 2350 U.T.

Geoc. Libr. { l = +5°.71
for 24ʰ { b = +4°.28

⊙'s Col. 99°.43 - 99°.93

Sel. Lat. +1°.29

Lighting & librational factors very favourable.

Mare Humboldtianum and Belkovitch

The Mare was called appropriately Humboldtianum after the explorer Alexander von Humboldt whose discoveries linked the Earth's eastern and western hemispheres. This 'sea' links the near and averted sides of the Moon.

It is only seen in its entirety at those libratory conditions which favour the NE limb and, of course, for the effective examination of its topographical features, the lighting has to be suitable at such times.

Conditions on 1986 August 20 provided an opportunity to study the elevations marking the eastern border of this small lunar basin. These took on the appearance under evening light of a series of mountain blocks, brilliantly lit, and standing out of the deepening shade at their bases. Detail on the Mare surface was not attempted on this occasion as the lighting was insufficiently advanced, but the general outline of Humboldtianum and the positions of neighbouring formations were indicated on the drawing.

To the north, the intruding walled-plain Belkovich, whose mean position lies at 62° N 90° E (i.e. on the mean limb) was well presented under the strong northerly libration. Interesting features caught at this particular stage of illumination were the *grey* penumbral bands where shadows from the three central elevations within Belkovich were about to be thrown fully on to the eastern wall.

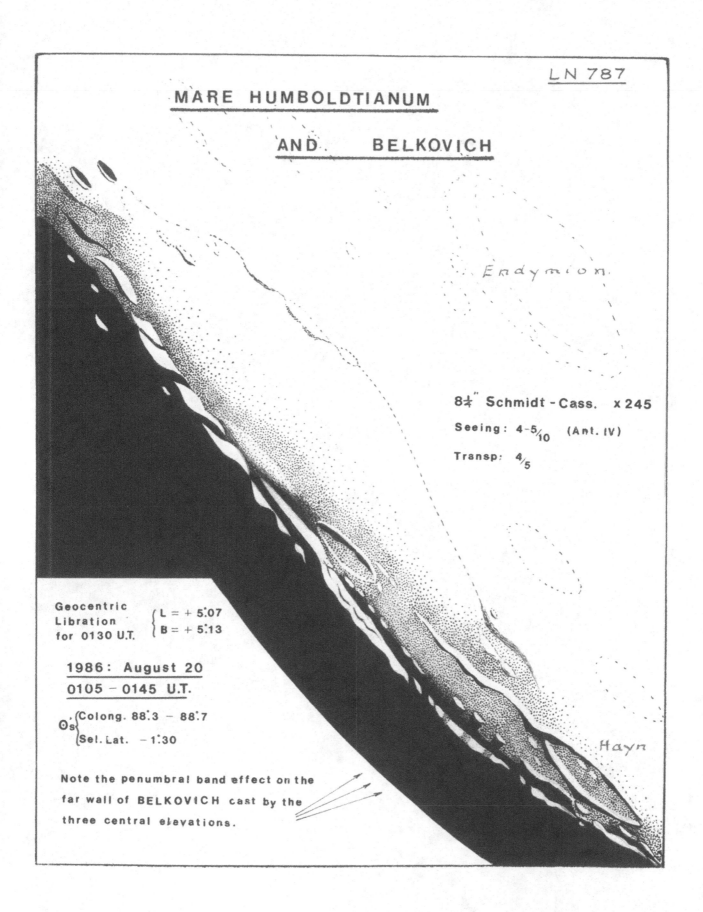

LN 787

MARE HUMBOLDTIANUM

AND BELKOVICH

Endymion

8¼" Schmidt - Cass.   x 245

Seeing: 4-5/10   (Ant. IV)

Transp: 4/5

Geocentric
Libration
for 0130 U.T.
$\{$ L = + 5°.07
  B = + 5°.13

1986: August 20

0105 - 0145 U.T.

Os $\{$ Colong. 88°.3 – 88°.7
     Sel. Lat. – 1°.30

Note the penumbral band effect on the
far wall of BELKOVICH cast by the
three central elevations.

Hayn

Quadrant II – Section 5

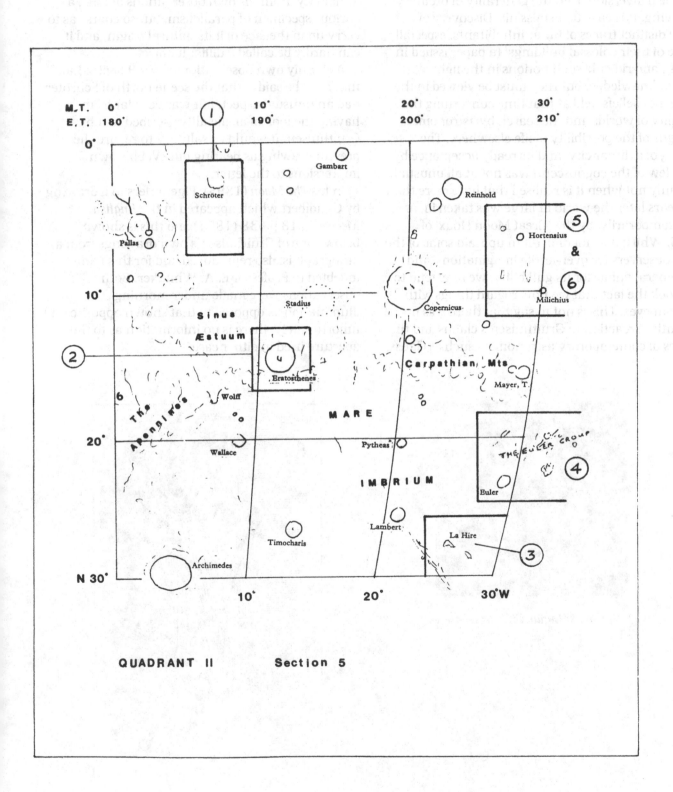

| M.T. | 0° | 10° | 20° | 30° |
| E.T. | 180° | 190° | 200° | 210° |

0°

Schröter ①

Gambart

Reinhold

Pallas

Hortensius ⑤ & ⑥

Milichius

10°

Stadius

Copernicus

Sinus Æstuum

②

Eratosthenes ④

Carpathian Mts

Mayer, T.

The Apennines

Wolff

MARE

THE EULER GROUP

20°

Wallace

Pytheas

IMBRIUM

Euler ④

La Hire

③

Lambert

Timocharis

N 30°

Archimedes

10°   20°   30°W

QUADRANT II   Section 5

45

Gruithuisen's 'Lunar City'

Consideration of the somewhat outré observations of Baron Franz von Paula Gruithuisen, 1774–1852 – a German physician and astronomer – will always provide a diversion from the generality of ordinary observing. It is easy to dismiss his 'Discovery of many distinct traces of lunar inhabitants, especially of one of their colossal buildings' (a paper issued in 1824), and ridicule such notions in the light of present knowledge, but they must be viewed in the context of beliefs held at that time concerning the plurality of worlds and the search by astronomers for signs of the possibility of life elsewhere. The finding of a 'lunar city' and its ready acceptance by not a few of the cognoscenti was not at all unusual, certainly not when it is realised that little more than ten years later the world at large was taken in, at least temporarily, by the 'Great Moon Hoax' of 1835. Whilst it is not difficult to upbraid some of the early observers for over-fertile imaginations, and the general public for its gullibility, we may tend to overlook the fact that we now regard things with different eyes. This is not to say that there was no dissentient reaction to Gruithuisen's claims and the names of contemporary astronomers such as Gauss,

Littrow and Olbers can be cited in this connection – although guarded pluralists themselves.

Much later, T. W. Webb described the fabled 'lunar city' from his own observations as: '. . . a curious specimen of parallelism, but so coarse as to carry upon the face of it its natural origin, and it can hardly be called a difficult object'.

As for my own observation of 1989 Sept.23 all that need be said is that the scene north of Schröter had an unusual aspect, but was very far from having the look of artificiality ascribed to it by Gruithuisen. It would be salutary to regard the present drawing as bearing out Webb's own impressions to the letter.

In his *The Moon* (1895), Elger refers to a drawing by Gaudibert which appeared in the *English Mechanic*, **18** p.638 (1874) and this is shown below. One of Gruithuisen's own drawings from a lithograph is also reproduced and for this I am indebted to Professor E. A. Whitaker. Both observations were made under morning illumination as opposed to that shown opposite and, unfortunately, there is no information as to the aperture used in either case.

Gruithuisen's 'Lunar City' as he drew it.

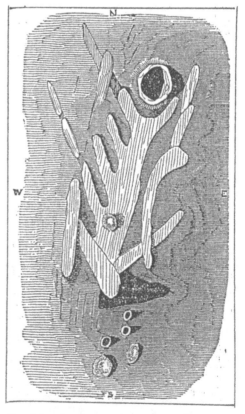

Gaudibert's rendition.

46

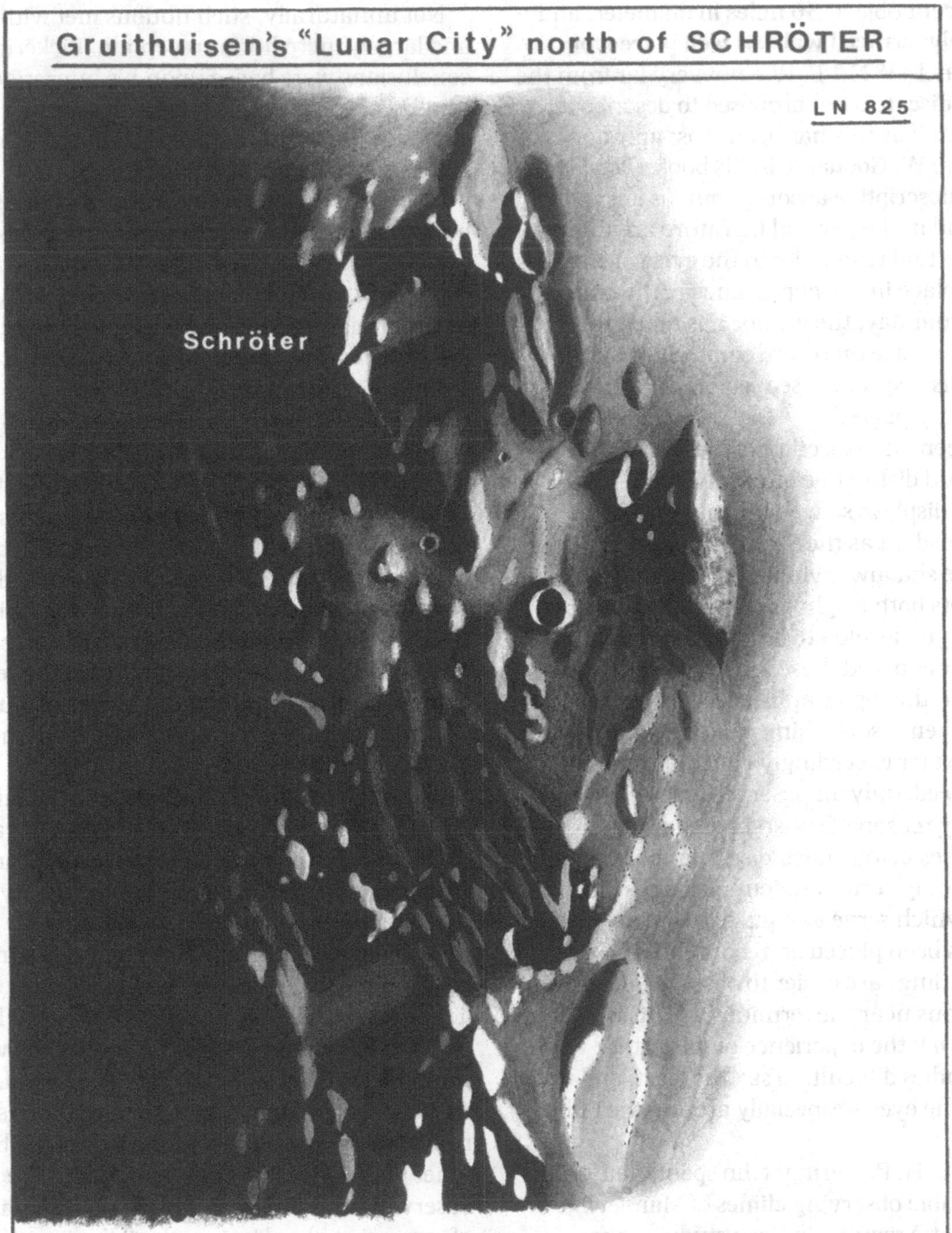

Gruithuisen's "Lunar City" north of SCHRÖTER

LN 825

Schröter

8¼" Schmidt-Cass. × 245
Seeing: 6-7/10   Transp: 5/5

1989 September 23
0245 - 0400 U.T.
⊙'s Colong. 185°.15 - 185°.80

Eratosthenes

This magnificent object, 36 miles in diameter, and with massively terraced walls, is well placed for observation at $14\frac{1}{2}°$ N $11°$ W – not very far from the centre of the disc. It is not proposed to describe its features in detail as this has been most ably done elsewhere, e.g. W. Goodacre in his book *The Moon* gives a good descriptive account, but, as it is not possible to find in the general literature anything other than textual references to the cycle of changes which takes place in the appearance of the crater during the lunar day, the author has taken this opportunity to place on record some of his own findings in this respect . . . see sequence of drawings on the following pages.

It will be seen on inspection of these that the sharp relief and distinctive outlines which Eratosthenes displays so well under a low Sun become less evident as the Sun's altitude increases. As discernible shadow dwindles, a system of dusky smudges, spots both bright and dark, and linear features begin to develop to become very prominent around local noon and these, having no apparent relationship to the topographical relief of the crater, make Eratosthenes something of an enigma, besides making the crater exceedingly difficult to trace at this time. Indeed, only an observer's close familiarity (by reason of constant study) with the metamorphoses being enacted enables him to positively identify certain salient markings and alignments which serve as a guide to the area under survey. It has been placed on record that these high-Sun shadings are easier to observe and draw than formations near the terminator. This is emphatically *not* the experience of the author who finds such studies difficult, to say the least, and very trying upon the eyes – especially around the Full Moon period.

Professor W. H. Pickering, who spent many years in the favourable observing climes of Mandeville, Jamaica and at Arequipa in Peru studying the behaviour of these markings, maintained that though the distribution of the general pattern of dark areas went through a predictable cycle from lunation to lunation, it was not so in the case of the smaller details which he found subject to the most capricious changes, and this led him to the belief that they were due to migratory movements of 'swarms of small animals or insects' over the relatively short season of the lunar day. Equally fanciful was his reference to the area as 'The Gardens of Eratosthenes'!

Not unnaturally, such notions met with a mixed and largely incredulous reception. Pickering's conclusions were based upon his interpretation of small-scale details which were of so subjective a character that another observer might draw and interpret them quite differently.

The series published herewith has been selected from drawings made during an early period of the writer's observing career and have been chosen to show the changes at approximately $12°$ intervals in the illumination. Unfortunately, the sequence is incomplete as a number of post-Full Moon drawings were lost in transit during exchange communications with a colleague in the United State – long before the time of effective photocopying! Later observations in the 1970s and 1980s were made with larger instruments and have not been included because of lack of aperture uniformity. It is to be hoped that interested amateurs will take up and continue what is a fascinating, if demanding, branch of lunar study, as there is still much to be learned from the reaction of the lunar 'soil' to changing angles of illumination – as distinct from the general concept of the 'incident-light' theory.

Although but few Transient Lunar Phenomena (TLP) events have been vouchsafed to the author in his observing, mention ought to be made in dealing with Eratosthenes of an anomalous observation on 1947 January 30 (mean colongitude $16°$) in which the main peak of the massive central mountain group appeared to be in a shadowless condition at a time when, having regard to its claimed height of 6600 ft, the whole area of the floor to the west should have been still in darkness. Instead, immediately to the west was a dark (intensity $1\frac{1}{2}$–2) region extending almost to the foot of the bright inner wall and very diffuse in outline. The observation could not be followed through because of increasing cloud but on the following night all was normal. Unavoidably, we have to enter the realm of speculation as to the cause of this singular appearance and one can only infer the partial obscuration of the normal shadow by overlying dust thrown to a considerable altitude either by a meteoric impact or an endogenous eruption. Confirmation of this event was sought at the time but was not forthcoming.

(1) A sequence of the changing appearances of
ERATOSTHENES throughout the lunar day

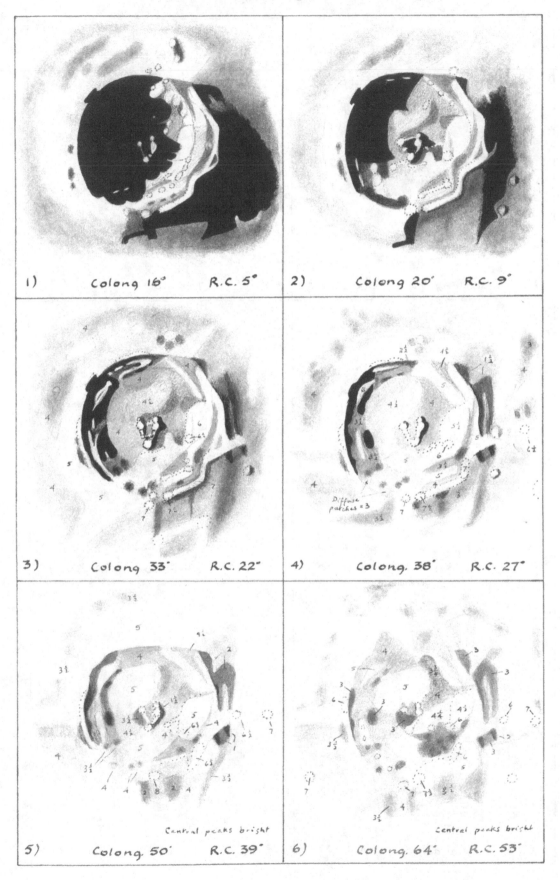

**(2) A sequence of the changing appearances of
ERATOSTHENES throughout the lunar day**

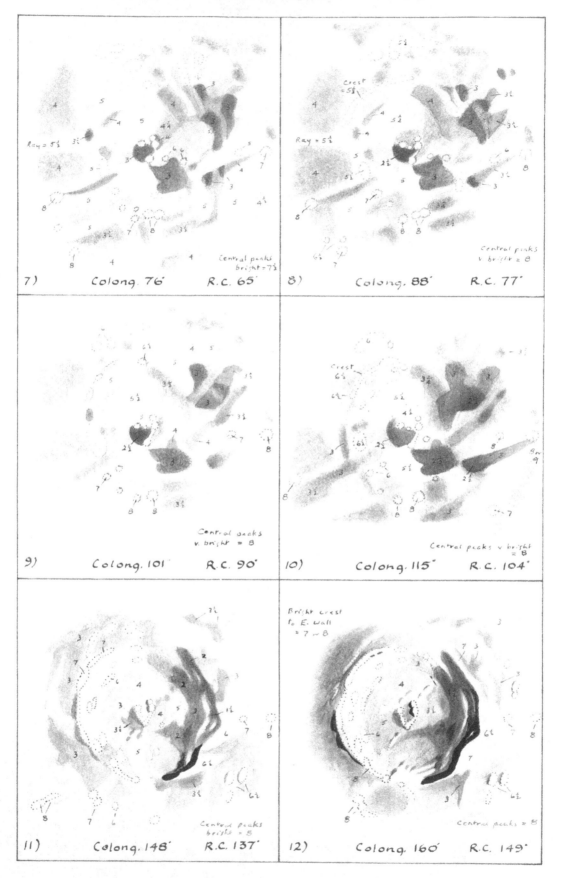

7)   Colong. 76°   R.C. 65°

8)   Colong. 88°   R.C. 77°

9)   Colong. 101°   R.C. 90°

10)   Colong. 115°   R.C. 104°

11)   Colong. 148°   R.C. 137°

12)   Colong. 160°   R.C. 149°

The La Hire Region

Mountain blocks which occur in isolation on a marial surface are often extremely bright, especially when observed at or near the terminator. The brightness would seem to be intrinsic in the majority of cases and not merely an effect of contrast against a dark background. La Hire is one such example; others are Piton, Pico, members of the Harbinger Mts, the multi-peaked mass situated WSW of Euler, and many more. They would appear to consist of material more resistant to laval flooding or other erosive agencies than the majority of features found in these vast basins. In some instances there is evidence that such elevations are the remains of sub-structures overwhelmed by marial inundation.

It is possible that La Hire and its small companion mt. group to the NW may have been more closely linked in the past; high-Sun studies may help to determine if such was indeed the case. As seen under low evening light (see drawing) a broad darkish band connects the two and observations conducted at a later, more critical, stage might establish whether this appearance indicates darker material or an actual slump in the mare surface at this site. On this occasion, the band was seen to be crossed by bright lines indicative of narrow rilles or possibly fault faces. Previous observations had given no sign of these features but on this occasion they were glimpsed repeatedly in the best moments.

The grand spire of shadow which La Hire casts when near the terminator may suggest that the mountain has a sharp peaked profile. Appearances can be deceptive because, whilst the base is some 12 miles in length, its height is rather less than a mile. It is also good to bear in mind how little, even in the best seeing, we are able to make out in terrestrial terms of the smallest visible detail. The distance thrown by the spire in the present instance is some 40 miles and the smallest feature recorded about 1 mile across. The linear features, though somewhat less than this in width, are discernible by reason of their length.

La Hire is sometimes described as having a summit craterlet; there is a comparatively easy one on the east-facing slope (hidden by shadow in the drawing) and another, more suspect crater on the NW side which may be, in reality, merely a bay between two spurs which contains shadow at sunset. Again, aspect can be deceptive and study under all stages of lighting is necessary.

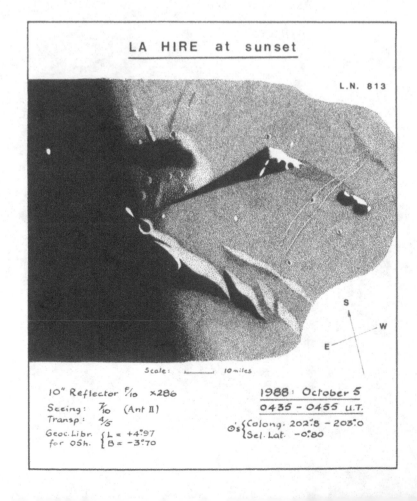

LA HIRE at sunset

L.N. 813

S
W
E

Scale: ⊢———⊣ 10 miles

10" Reflector F/10 ×286

Seeing: 7/10 (Ant II)
Transp: 4/5
Geoc. Libr. { L = +4°.97
for 05h. { B = -3°.70

1988: October 5
0435 – 0455 U.T.

⊙'s { Colong. 202°.8 – 203°.0
{ Sel. Lat. -0°.80

The Euler Group

To the S and SW of the crater Euler lies an extensive and rather scattered group of subjects. The disposition of the isolated hills and hillocks are rather reminiscent of the members of a megalithic circle particularly when these are seen near the terminator, at which times they throw long tapering shadow profiles. Some of the hills are rounded, others conical, and rather suggestive of volcanoes yet under the excellent conditions experienced on 1988 April 26 no summit craterpits could be detected despite the utmost attention. However, an elongated dome-like elevation, somewhat lemon-shaped, showed a distinct craterlet on its summit – the position of this dome is indicated by arrows. Another feature of interest was the undesignated mountain situated to the SE of the general assemblage; this appeared to have a half-submerged craterlet on its lower SE flanks, giving the strong impression that scree or rubble from the mountain has partially covered it (also arrowed).

The largest crater of the Euler group is the shallow ring P, some 7 miles in diameter, with low walls.

It is curious that so prominent an object as the multi-peaked mountain mass, lying some distance WSW of Euler, should have received no name on modern maps as it has a most imposing aspect at this stage of lighting. In Neison's nineteenth century chart, however, it is designated Euler beta and Wilkins followed him in retaining this.

There would appear to be few, if any, drawings extant of this region, so evidently it received little attention from amateur lunarians – at least to the extent of making a representation.

Under a high Sun, many members of the assemblage appear as bright points of light.

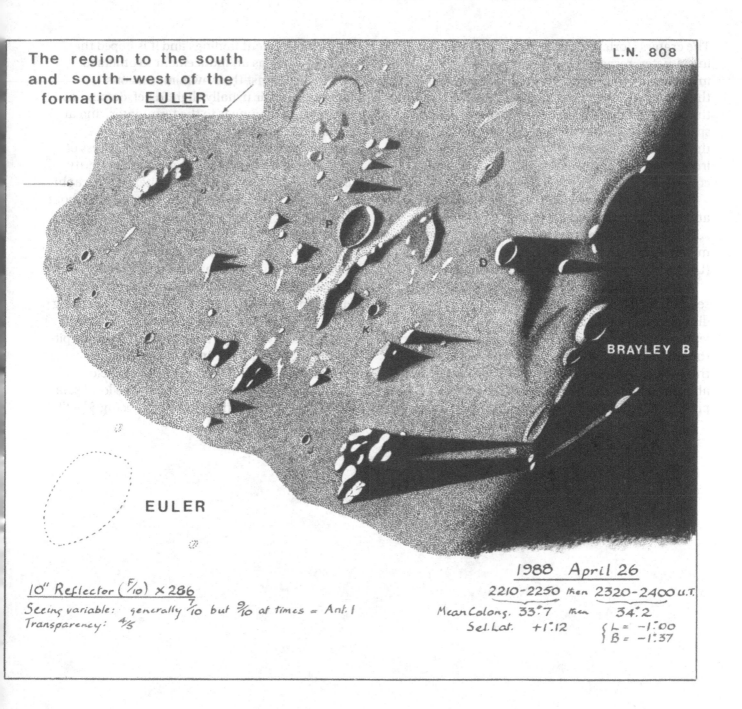

The region to the south and south-west of the formation EULER

L.N. 808

BRAYLEY B

EULER

10" Reflector (F/10) ×286
Seeing variable: generally 7/10 but 9/10 at times = Ant. I
Transparency: 4/5

1988 April 26
2210-2250 then 2320-2400 U.T.
Mean Colong. 33°.7 then 34°.2
Sel. Lat. +1°.12
L = -1°.00
B = -1°.37

The Hortensius – Milichius Region

The craters Hortensius and Milichius, 9 and 8 miles in diameter, respectively, appear rather insignificant on the chart but the area surrounding these formations and also around Tobias Mayer to the NE, is fascinating for those observers whose special interest is the study of domes, because here there is an assemblage of various types ranging from the classical concept to the irregular and complex.

This region received a certain amount of attention from the mid-1930s onwards and subsequently formed part of an intensive survey by members of both the BAA and ALPO lunar sections (UK and USA organisations respectively).

It is strange that observers' records which span several decades should continue to show disagreement in the interpretation of the smaller features associated with the domes, and even over the shapes and characteristics of some of the domes themselves. Although these differences have been ably discussed elsewhere, the drawings here presented which cover regions (5) and (6), show the writer's more recent findings and it is hoped that these may serve as an inducement to others to check them against their own impressions.

Because of their usually slight relief, domes have to be studied when near to the terminator, and at solar altitudes from 5° to 10° they become increasingly difficult to trace. Greatly in excess of 10°, they are lost to view entirely in the majority of cases and in this respect the Orbiter photographic records are deficient in providing information about such elusive or transient appearances. As a result, opportunities remain in making careful topographical studies and for adding to our still limited knowledge of such structures.

The author has found from experience that in addition to adequate aperture, appropriate lighting and good seeing, an equally important factor in securing effective results is *air clarity*. The craterpits marking the summits of five domes in the observation of 1987 November 30 were held steadily during the better moments – the local solar altitude for those north of Hortensius being $3\frac{1}{2}$–4°.

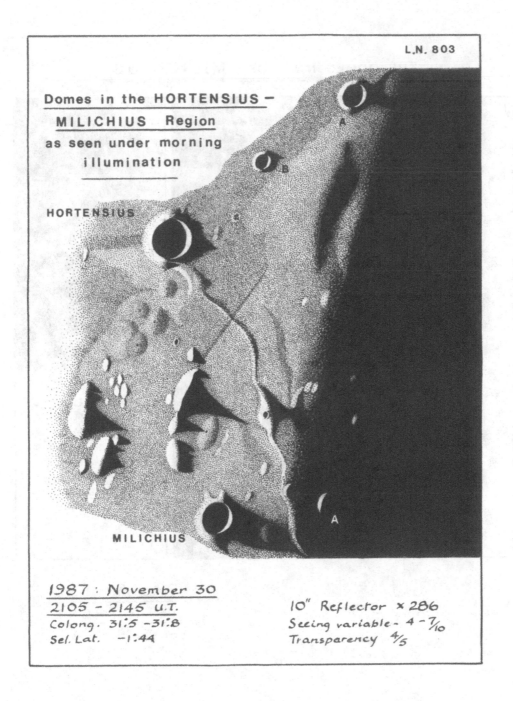

L.N. 803

Domes in the HORTENSIUS –
MILICHIUS Region
as seen under morning
illumination

HORTENSIUS

MILICHIUS

1987 : November 30
2105 – 2145 U.T.
Colong. 31°5 – 31°8
Sel. Lat. –1°44

10" Reflector × 286
Seeing variable – 4 – 7/10
Transparency 4/5

THE DOMES NNW OF   MILICHIUS

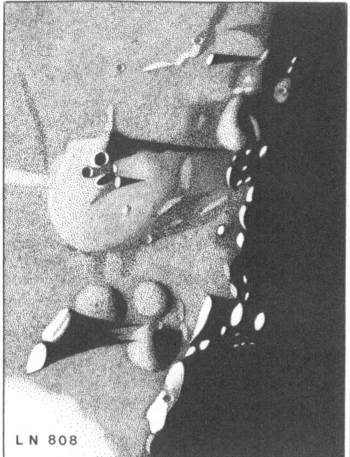

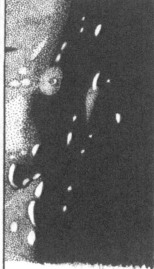

LN 808

A localised sketch of
the region to the S.W
at 2205 (Colong. 33°.5)
confirming the appearance
of a craterlet on a
mound or dome (?)
under somewhat
higher lighting.

Harold Hill.

10" Reflector (F/10) × 286
Seeing 6-8/10 (Ant II)
Transp. 4/5

Geoc. Libr. {L = -0°.49
for 21ʰ UT. {B = -1°.86

1988 April 26
2055 - 2140 U.T.
⊙ {Col. 32°.9 - 35°.3
  {Sel. Lat. +1°.12

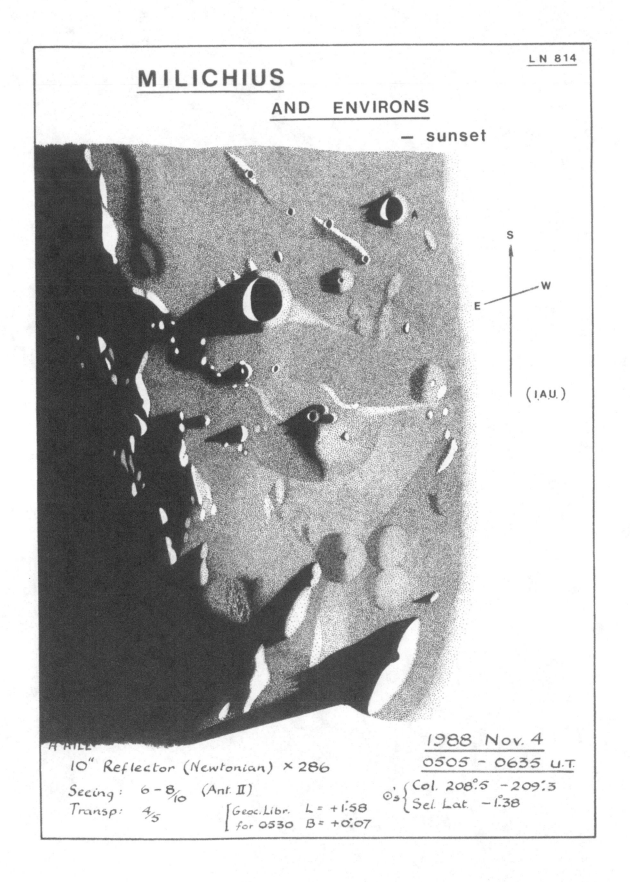

LN 814

MILICHIUS

AND ENVIRONS

— sunset

H. HILL

10" Reflector (Newtonian) × 286

Seeing: 6 - 8/10 (Ant. II)

Transp: 4/5

Geoc. Libr.   L = +1.58
for 0530   B = +0.07

1988 Nov. 4

0505 - 0635 U.T.

☉'s { Col. 208.°5 - 209.°3
      { Sel. Lat. -1.°38

S
W
E
(I.A.U.)

Quadrant II – Section 6

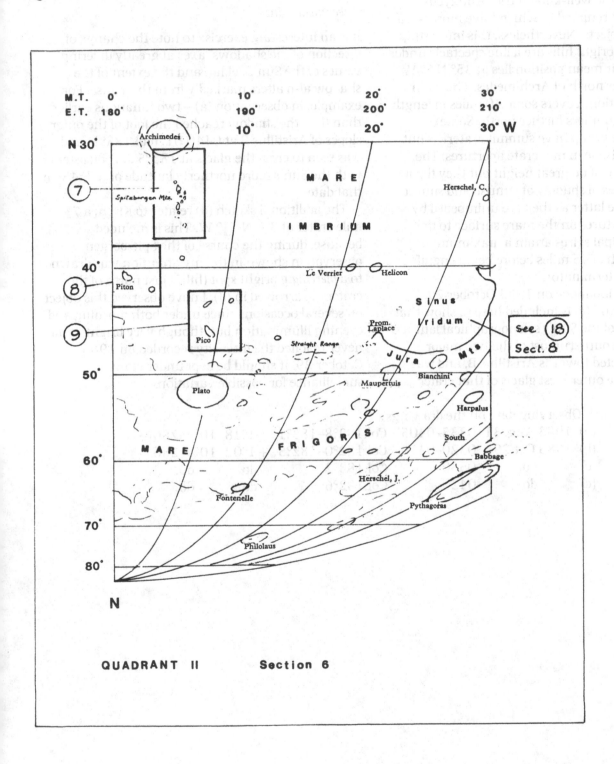

QUADRANT II   Section 6

The Spitzbergen Mountains

The Spitzbergens have not received the close attention they deserve from lunarians probably because of the presence of such imposing neighbours as the well-known trio, Autolycus, Aristillus and Archimedes, which have more to offer as telescopic objects. Nevertheless, this interesting group of small bright hills are a fine spectacle under a low Sun; their mean position lies at 35° N 5° W – some 75 miles north of Archimedes. The main cluster of elevations covers some 40 miles in length from N–S with outliers further north. Some members of the group have summit craterlets but these are not visible in moderate apertures. The Spitzbergens are of no great height but they throw impressive spires of shadow at sunrise and sunset – especially at the latter as they are unimpeded by any sizeable features on the mare surface to the east. The principal spires attain a maximum projected length of 45 miles before being engulfed by the evening terminator.

The observation made on 1983 October 29, depicted in (b) and (c) is included here to show that with the value of the Sun's selenographical latitude at + 1°02, the southernmost of the two major spires was directed towards Aristillus B, and was seen to cross the outer west glacis of that crater at colongitude 183°2 as a grey penumbral shade – see drawing (c).

A technical point

It is an interesting exercise to note the change of direction of the shadows' axes at greatly differing values of the Sun's sel.lat. and the extent of the shadow also alters markedly from this cause. For example, in observation (a) – two lunations earlier than (b) – the shadow reached the foot of the outer slopes of Aristillus B at 0440 UT. (col.183°3) and was seen to cross the glacis at 183°5 . . . consistent with the Sun's more northerly latitude of + 1°48 on that date.

The additional sketch (d) relates to Kirch, a 7¼ mile crater at 39° N 5½° W. This is included because, during the course of the Spitzbergen observation shown in (b), my attention was drawn to a *piercingly* bright spot (hill?) just north of the crater . . . arrowed in (d). I have observed this object on several occasions since under both morning and evening illumination but, though very bright, it has never reached the intensity recorded on 1983 October 29. It should be kept under close surveillance for possible variations.

Observing data on the drawings:

(a)	1983 Aug. 31	0335–0405	Col.182°8–183°1	+1°48	10″ × 286
(b)	1983 Oct.29	0320–0410	Col.182°0–182°5	+1°02	10″ × 286
(c)	do.	0530	Col.183°2	do.	do.
(c)	do.	0420	Col.182°6	do.	do.

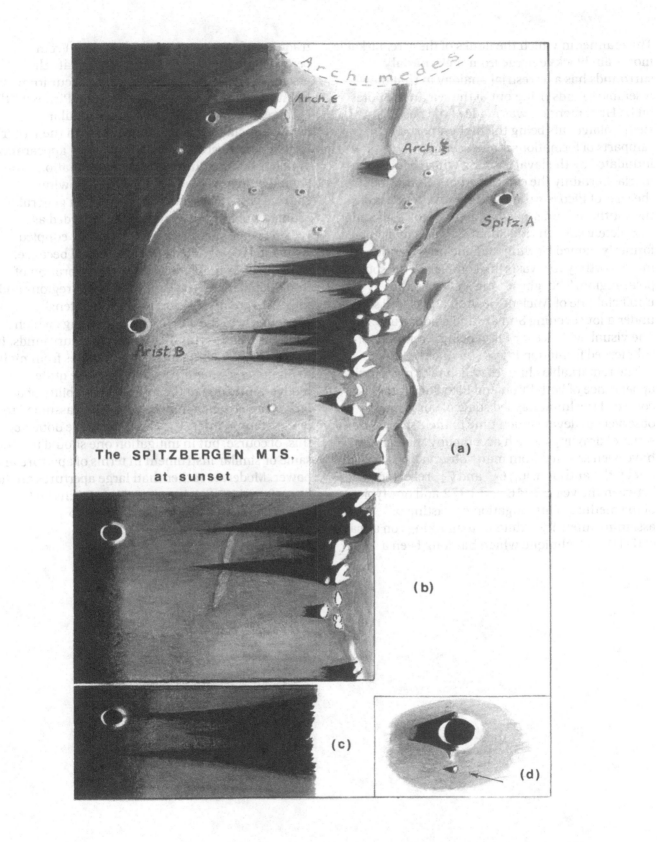

The SPITZBERGEN MTS,
at sunset

The isolated Mountains Piton and Pico

The manner in which the flanks of these remarkable mountain blocks emerge from their marial surrounds has a terrestrial analogy in that they resemble islands rising out of the sea. In the notes on La Hire, mention was made of the possibility of such isolated mts being the highest portions of ramparts of formations almost completely inundated by the laval flooding which caused the maria. Certainly the evidence that this may be so in the case of Pico is strongly suggestive as it lies on the southern border of what appears to be an almost completely sunken ring about 75 miles in diameter, formerly named Newton, a name now transferred more worthily to a vast ring complex in the south polar region. The 'ghost' ring still bears the unofficial title of Ancient Newton and is well seen under a low evening Sun around colongitude 184°. The visual evidence for Piton being a remnant of a submerged formation is less convincing.

The remarkable changes which take place in the appearance of both Piton and Pico during the course of the lunar day led some of our keenest observers to devote much time to their study. The series of drawings which accompany these notes have been selected from many observations carried out by the author using $6\frac{1}{2}''$ and $7\frac{1}{4}''$ reflectors between the years 1946 and 1972 and one typical of the method of investigation consisting of assigning intensity estimates to markings on a scale of 0–10 – a technique which has long been a tradition of American lunar observing. From personal results it seems difficult to equate the essentially repetitive changes which occur from lunation to lunation to both Piton and Pico with the early reports of 'snowstorms' and irregular behaviour claimed from time to time. In the writer's experience, most so-called 'anomalous' appearances which occur are due to errors in observation. One should be ever vigilant against the following sources of such observational errors. In general, estimates of intensity can only be regarded as reliable when made in a clear, dark sky coupled with good seeing. The latter is essential because, under poor or indifferent seeing, the vibration of lunar detail mixes the light of adjacent regions and may thus create false impressions of intensity – particularly in the case of small markings which contrast greatly with their immediate surrounds. In the course of making intensity estimates from night to night, *large-scale* errors may be made quite inadvertently due to the inherent possibility of a lack of areas which may be admissibly assumed to be constant in intensity. Nothing can be done about this, of course, but in mitigation one should use the same or similar instrument in terms of aperture and power. Moderate rather than large apertures are to be preferred in such investigations, i.e. in the 6″–12″ range as they are less susceptible to atmospheric vagaries.

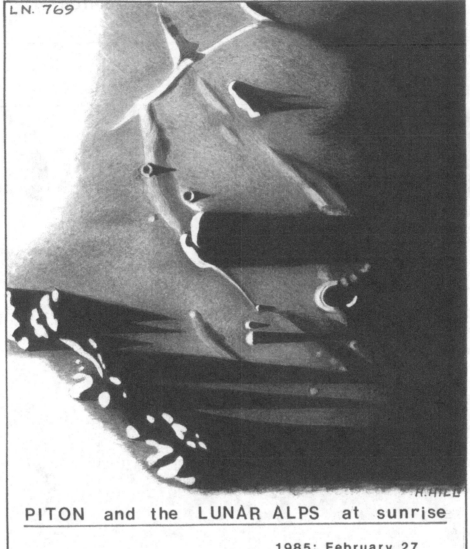

LN. 769

PITON and the LUNAR ALPS at sunrise

1985: February 27

1955 – 2020 U.T.

10" Reflector x190

S : 5 (Ant.III)   T : 4

Sun's Colong. 4°.44   4°.65

Sel.Lat.   −1°.50

A sequence of the changing appearances of Mt. PITON
throughout the lunar day

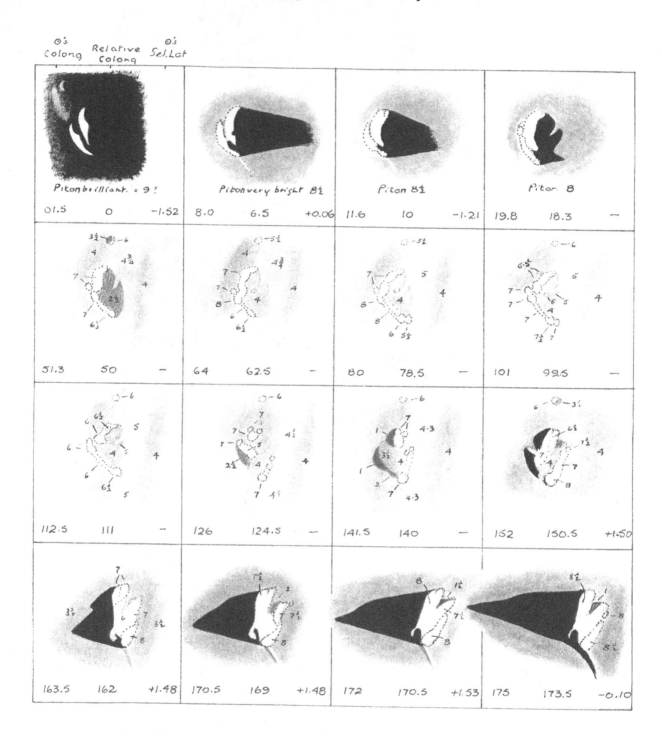

⊙'s Colong   Relative Colong   ⊙'s Sel. Lat

Piton brilliant. = 9?

01.5   0   −1.52

Piton very bright 8½

8.0   6.5   +0.06

Piton 8½

11.6   10   −1.21

Piton. 8

19.8   18.3   —

51.3   50   —

64   62.5   —

80   78.5   —

101   99.5   —

112.5   111   —

126   124.5   —

141.5   140   —

152   150.5   +1.50

163.5   162   +1.48

170.5   169   +1.48

172   170.5   +1.53

175   173.5   −0.10

PICO & PICO B. *(sunset)*

LN. 318

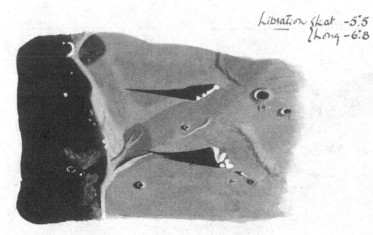

Libration {Lat -5°.5
{Long -6°.8

Instrument.
 6½" Reflector x220

1948 Sept. 25
13ᵇ 05ᵐ to 13ᵇ 45ᵐ G.M.T.
Col. 184°:4 h 184°:8

{Seeing 5/10
{Transp. 4/6.

Note   Precise positions of tips of the shadows noted at 13ᵇ 05ᵐ before details drawn. 40 min later the shadow of Pico was crossing the forked ridge.

A sequence of the changing appearances of Mt. PICO
throughout the lunar day

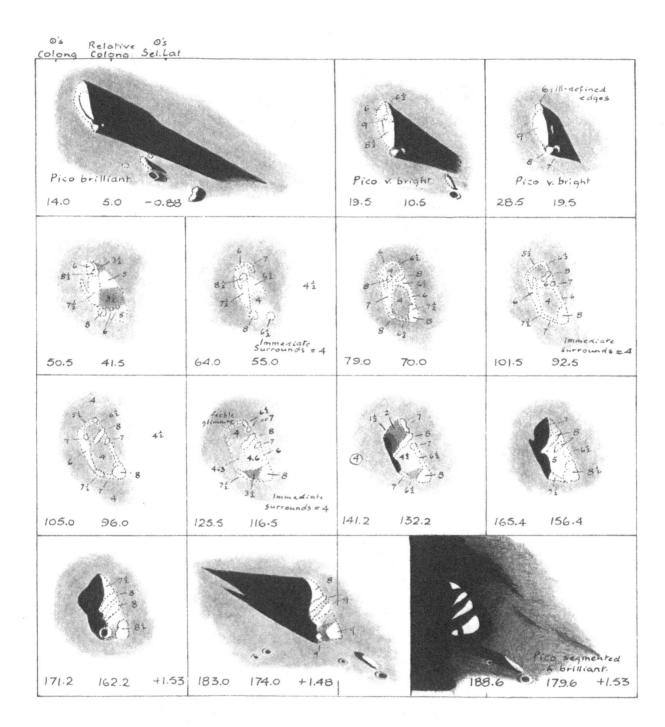

Quadrant II – Section 7

Quadrant II – Section 7

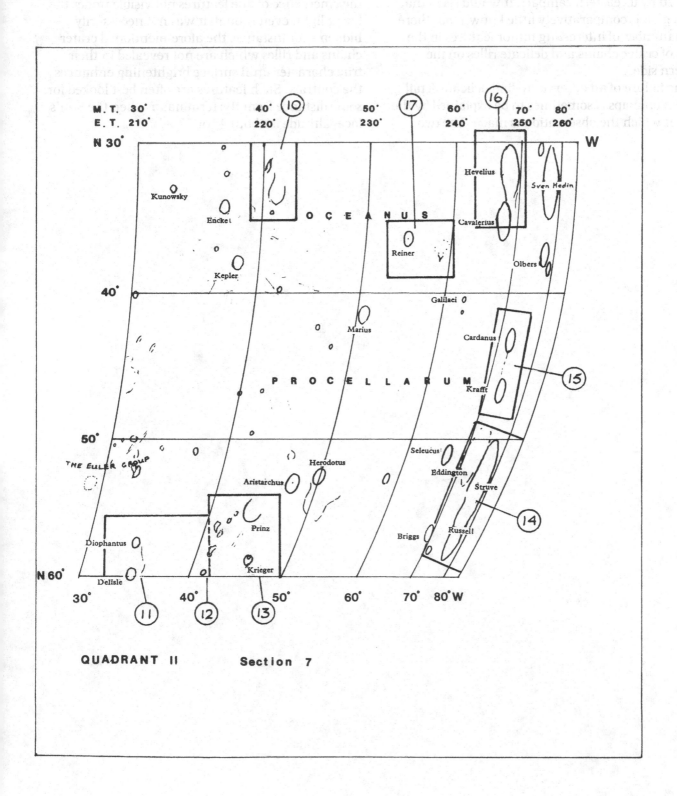

QUADRANT II   Section 7

The region SW of Encke

The ruined walled-plain called Maestlin R presents a somewhat brief but dramatic scene soon after sunrise by reason of the fine spires of shadow thrown by its eastern rampart. It would seem that the region is comparatively little known, but there are a number of interesting minor features in the form of crater chains and delicate rilles on the eastern side.

The failure of an observer to find delicate detail shown on maps is sometimes to be explained by the *time* at which the observation is made. The two observations shown are morning aspects made 4 hours apart and quite independently of each other. They show the rapid retreat of shadow and the emergence of fine features not visible under the lower light, even though it was not necessarily hidden – for instance, the afore-mentioned crater chains and rilles which are not revealed in their true character until surface brightening enhances the contrast. Such features are often best looked for some distance from the terminator when the Sun's local altitude is about 4° or 5°.

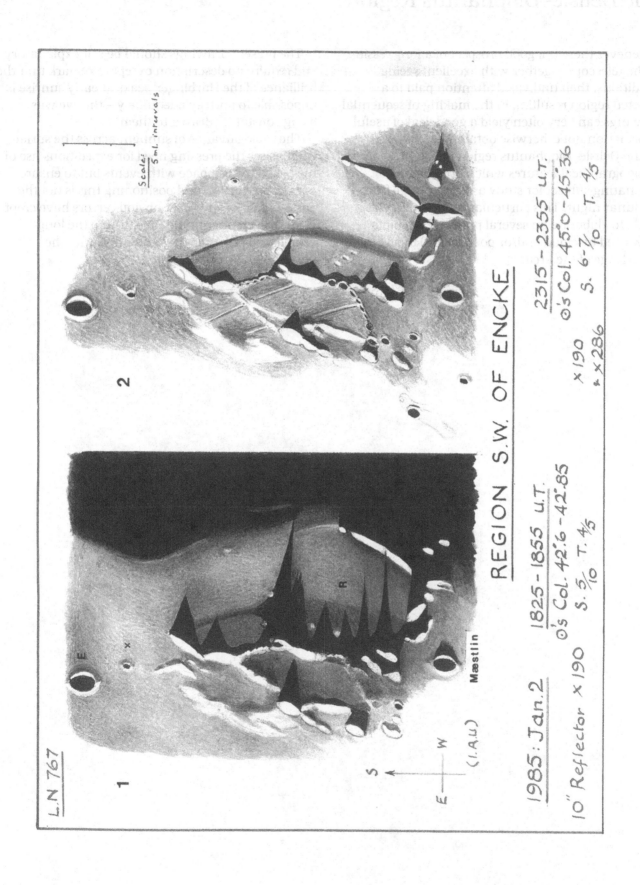

L.N 767

1

2

Scale: / 5 ml. intervals

Maestlin

E
W
(I.A.U.)
S

REGION S.W. OF ENCKE

1985: Jan.2   1825-1855 u.T.
⊙'s Col. 42°.6 -42°.85
S. 5/10   T. 4/5

10" Reflector ×190

2315-2355 u.T.
⊙'s Col. 45°.0 -45°.36
S. 6-7/10   T. 4/5

×190
* ×286

The Delisle–Diophantus Region

Whenever there is a good prospect of a long session at the telescope together with excellent seeing conditions, then undivided attention paid to a selected region resulting in the making of sequential drawings can very often yield a good deal of useful information not otherwise obtainable.

The Delisle–Diophantus region is rich in a variety of topographical features which provide a fascinating subject for study as they emerge from the lunar night. This particular observation was made to elaborate on several previous attempts (thwarted by cloud and/or poor seeing) to add to my knowledge of the area.

The present drawings should be self-explanatory and require no description except to remark that the brilliance of the Harbinger peaks at early sunrise is impossible to portray adequately – the eye was being constantly drawn to them.

The rapid advance of sunlight across the surface emphasises the pressing need for expeditious use of the pencil to keep pace with events but to ensure correct proportion and positioning this is not the easiest of tasks and some obvious errors have crept in. However, one should hold firm to the long-established principle that alteration *after* the observation must be resisted.

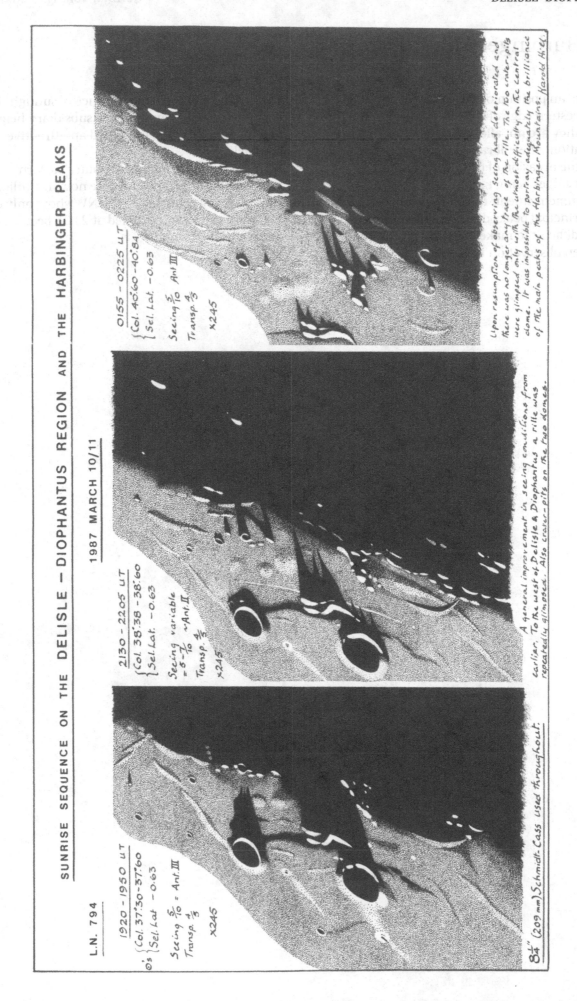

SUNRISE SEQUENCE ON THE DELISLE – DIOPHANTUS REGION AND THE HARBINGER PEAKS

1987 MARCH 10/11

L.N. 794

1920 – 1950 UT
⊙s { Col. 37°30′–37°60
{ Sel. Lat. –0.63

Seeing 5/10 = Ant. III
Transp. 4/5
×245

2130 – 2205 UT
{ Col. 38°38′–38°60
{ Sel. Lat. –0.63

Seeing variable
= 5–7/10 ~Ant. II
Transp. 4/5
×245

0155 – 0225 UT
{ Col. 40°60–40°84
{ Sel. Lat. –0.63

Seeing 5/10 Ant. III
Transp. 4/5
×245

84″ (209 mm) Schmidt-Cass used throughout.

A general improvement in seeing conditions from earlier. To the west of Delisle & Diophantus a rille was repeatedly glimpsed. Also crater-pits on the two domes.

Upon resumption of observing seeing had deteriorated and there was no longer any trace of the rille. The two crater-pits were glimpsed only with the utmost difficulty as the central dome. It was impossible to portray adequately the brilliance of the main peaks of the Harbinger Mountains. Harold Hill.

The Harbinger Mountains

As the mean position of this group of bright peaks on the western edge of the Mare Imbrium is $27\frac{1}{2}°$ N 41° W, they are not affected critically as regards presentation when libration is unfavourable as it was on the occasion of this observation. The relevant values being L$= + 3°.2$ B$= - 6°.0$.

At commencement, the coruscations undergone by the principal peaks due to indifferent seeing, made it difficult to make out their precise shapes, but, when observation was resumed later under improved conditions, the advance of sunlight had brought into view many of the subsidiary heights and the whole region presented an attractive spectacle.

In the second drawing, the outer eastern ramparts of the crater Prinz are now in sunlight but those of Krieger further to the NW show only dimly.

Observation was curtailed at 2105 because of the intervention of cloud.

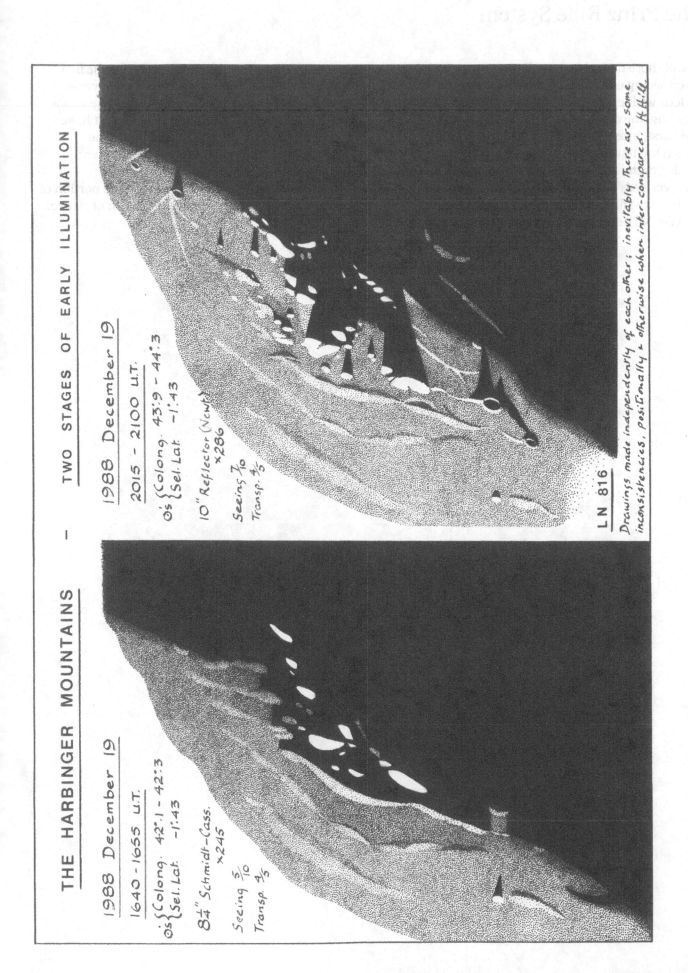

THE HARBINGER MOUNTAINS — TWO STAGES OF EARLY ILLUMINATION

1988 December 19

1640 - 1655 U.T.

$\odot$'s $\begin{cases} \text{Colong.} & 42°.1 - 42°.3 \\ \text{Sel. Lat.} & -1°.43 \end{cases}$

$8\frac{1}{4}$" Schmidt-Cass. ×245

Seeing 5/10
Transp. 4/5

1988 December 19

2015 - 2100 U.T.

$\odot$'s $\begin{cases} \text{Colong.} & 43°.9 - 44°.3 \\ \text{Sel. Lat.} & -1°.43 \end{cases}$

10" Reflector (Newt)
×286

Seeing 7/10
Transp. 4/5

LN 816

Drawings made independently of each other; inevitably there are some inconsistencies, positionally & otherwise when inter-compared. K. Hill.

The Prinz Rille System

One of those desirable, but all too rare, occasions when seeing conditions were above average for long periods with exceptionally sharp glimpses occurring enabling fine detail to be recorded without difficulty.

At first, attention was directed to the system of rilles lying to the north of the 32 mile diameter flooded crater Prinz – several of these rilles resemble dry river beds because of their winding courses.

The three finer rilles running approximately WSW–ENE nearer to the terminator, however, were of especial interest as these were seen for the first time. In addition, even more delicate transverse cross-members were detected. Indeed, there was a host of finer details hereabouts which must have been at, or beyond, the resolving limit of the telescope used, and no attempt at depiction or interpretation could be made.

Note the dome-like feature occupying a portion of the inundated interior of Prinz just south of centre.

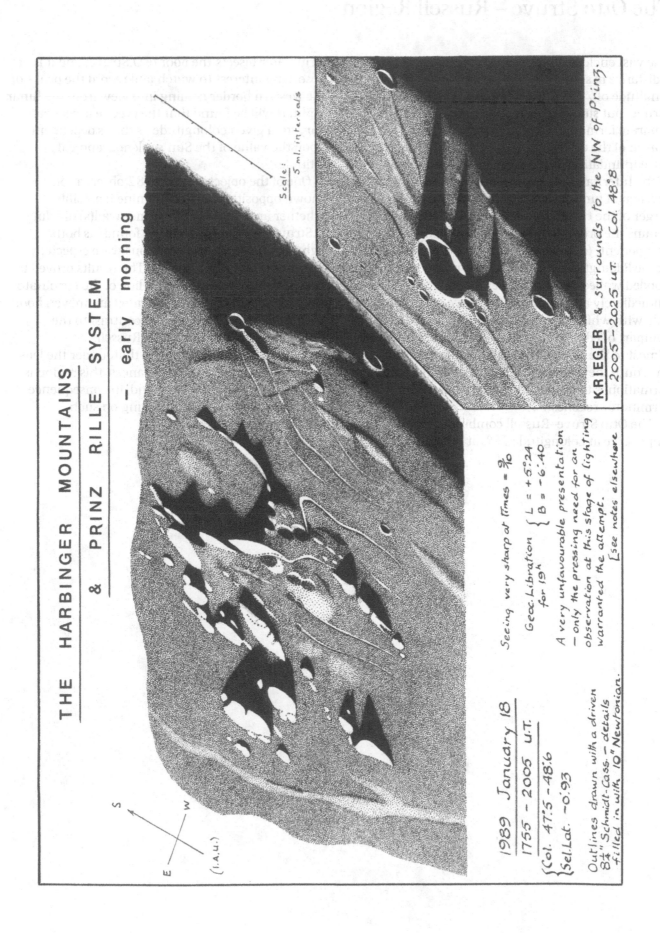

THE HARBINGER MOUNTAINS
& PRINZ RILLE SYSTEM
— early morning

1989 January 18
1755 – 2005 u.t.
{ Col. 47°.5 – 48°.6
{ Sel.Lat. –0°.93

Outlines drawn with a driven
8¼" Schmidt-Cass – details
filled in with 10" Newtonian.

Seeing very sharp at times = %

Geoc.Libration { L = +5°.24
for 19ʰ { B = –6°.40

A very unfavourable presentation
– only the pressing need for an
observation at this stage of lighting
warranted the attempt.
[See notes elsewhere.

S
W
E
(I.A.U.)

Scale:
5 ml. intervals

KRIEGER & surrounds to the NW of Prinz.
2005 – 2025 u.t.   Col. 48°.8

The Otto Struve – Russell Region

The vast enclosure stretching from 20° N–28½° N – a distance of some 175 miles – and at a mean longitude of 77° W, was formerly named Otto Struve, but since it consists in reality of two separate formations which have coalesced by reason of the destruction of the dividing walls by laval inundation, it was decided by the Commission of the International Astronomical Union, during the revision of lunar nomenclature, to redesignate. The larger of the two to the south, 114 miles diameter, retains the original name, but the northern component, 62 miles diameter, is now Russell. The name Eddington was given to the remains of a flooded walled-plain, 85 miles diameter, situated immediately adjacent to Otto Struve on its eastern side with which it shares a common wall. The E rampart of Eddington is much eroded and reduced almost to the level of the Oceanus Procellarum at the south, nevertheless, a fine view of this particular formation is to be gained when the morning terminator reaches 73°.

The Otto Struve–Russell combination is impressive at colongitude 78°, at which stage the terminator bisects the floor of O.Struve, and it is of absorbing interest to watch and record the peaks of the western border coming into view from the lunar night. It will be found that the precise aspect can vary for a given colongitude as this is dependent upon the value of the Sun's selenographical latitude.

One of the objects of the 1982 observation, shown opposite, was to determine if possible whether any traces of the former wall(s) dividing O.Struve and Russell could be found, as both lighting and libratory conditions were especially favourable for such a search. The results proved to be negative, but the irregular line of the terminator shading within O.Struve indicated an uneven floor with evidence of slumping, in contrast to the apparently smooth interior of Russell.

It may be of interest to note that, under the least favourable libratory circumstances, this region is removed almost to the limb and, in consequence, suffers from acute foreshortening on such occasions.

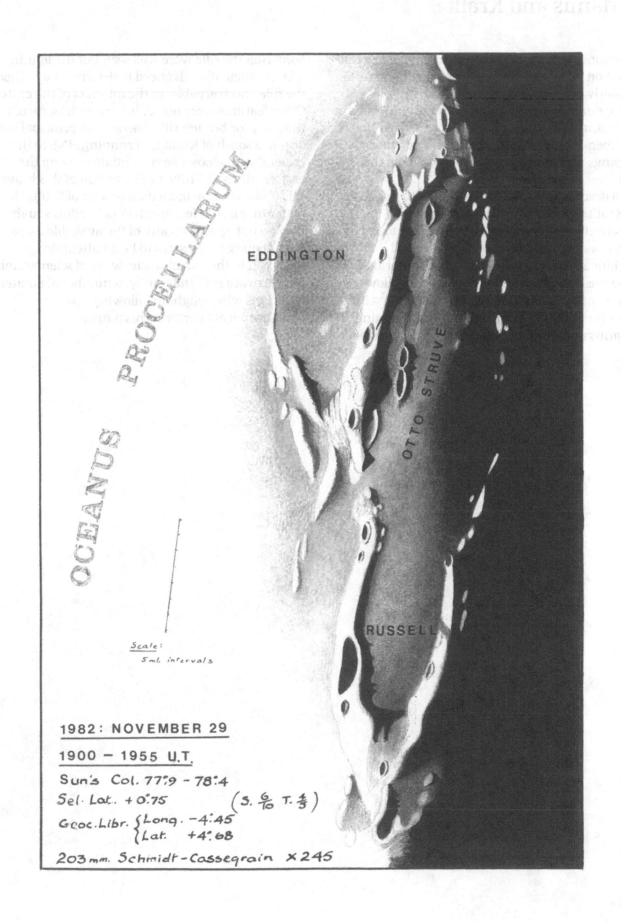

OCEANUS PROCELLARUM

EDDINGTON

OTTO STRUVE

RUSSELL

Scale:
5ml. intervals

1982: NOVEMBER 29

1900 – 1955 U.T.

Sun's Col. 77°9 – 78°4

Sel. Lat. +0°75   (S. 6/10 T. 4/5)

Geoc. Libr. { Long. –4°45
            { Lat. +4°68

203 mm. Schmidt-Cassegrain × 245

Cardanus and Krafft

One of those striking meridional arrangements that we find on the Moon; they would seem to occur too frequently to be regarded as merely fortuitous. Other examples to be found within Section 7 are the (Lohrmann)–Hevelius–Cavalerius trio, Sven Hedin and Olbers, Otto Struve and Russell and other groupings nearer to the limb not shown on the chart.

Full descriptive accounts dealing with Cardanus and Krafft (31 and 32 miles in diameter, respectively) and their environs are to be found in Elger, Goodacre and Wilkins & Moore – all descriptive guides to the Moon's surface formations.

The present observation was made to follow the progress of early illumination on the region with particular attention to the main rille connecting Cardanus and Krafft. At this stage, the raised banks bordering the rille were well seen but the lighting was insufficiently advanced to determine whether the rille was traceable on the interiors of the craters. Other features were noted, however, in particular a tributary, or branch rille, leaving the principal some distance south of Krafft and running NNE to the crater E with a possible continuation along the eastern glacis of Krafft. Confirmation of this branch rille and its course immediately west of E, together with other minor features NW of Krafft is sought as they do not appear on any of the available maps. Such requests may seem to be an attempted reversion to the old 'obsolete' ways of selenography but there are still, thankfully, a number of amateur lunarians who delight in following up cartographical queries of this nature.

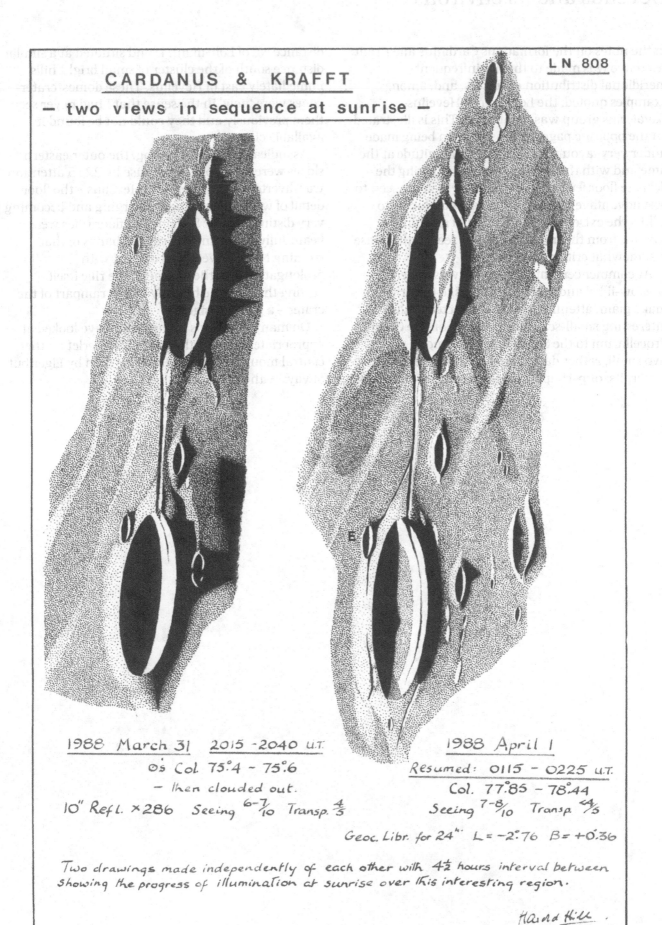

LN 808

CARDANUS & KRAFFT

— two views in sequence at sunrise —

E

1988 March 31   2015 - 2040 u.t.
O's Col. 75°.4 ~ 75°.6
— then clouded out.
10" Refl. ×286   Seeing $^{6-7}/_{10}$ Transp. $^4/_5$

1988 April 1
Resumed: 0115 - 0225 u.t.
Col. 77°85 - 78°44
Seeing $^{7-8}/_{10}$ Transp. $^4/_5$

Geoc. Libr. for 24ʰ L = -2°76 B = +0°36

Two drawings made independently of each other with 4½ hours interval between
showing the progress of illumination at sunrise over this interesting region.

Harold Hill.

Hevelius and its environs

In the notes on the formations Cardanus and Krafft, reference was made to the not infrequent meridional distribution of craters, and, among examples quoted, the Lohrmann–Hevelius–Cavalerius group was mentioned. This is illustrated on the opposite page, the observation being made under very favourable libration in longitude at the time and with the express object of studying the delicate floor features of Hevelius which are seen to best advantage if observations are conducted to follow the extremely rapid lifting of morning shadow from the interior. For this reason the phase is somewhat critical.

At commencement the interior was largely shadow-filled and, as the footnotes to the drawing make plain, attention was first directed to the interesting small-scale objects lying on the Oceanus Procellarum to the SE of Hevelius. Among these, two small, rather dark domes with summit craterpits, or perhaps crater-cones were found some

distance NE of Lohrmann D and situated at a similar distance south of the cluster of small bright hills immediately east of Hevelius. These domes/crater-cones were 'new' in the sense that I had never seen them previously and they could not be found in available charts.

As indicated on the drawing, the outer eastern slopes were not depicted because by 2215 attention was diverted to the interior of Hevelius – the floor detail of which was by now emerging and becoming very distinct. Under 8 seeing the floor rilles were beautifully sharp and the rugged banks of that running NE–SW were well defined with prolongation into the darkness, the rille itself cutting through the brightly lit SW rampart of the crater – a glorious sight.

On many occasions in the past I have looked at appropriate times for the summit craterlet on the central mountain within Hevelius seen by Elger but always without success.

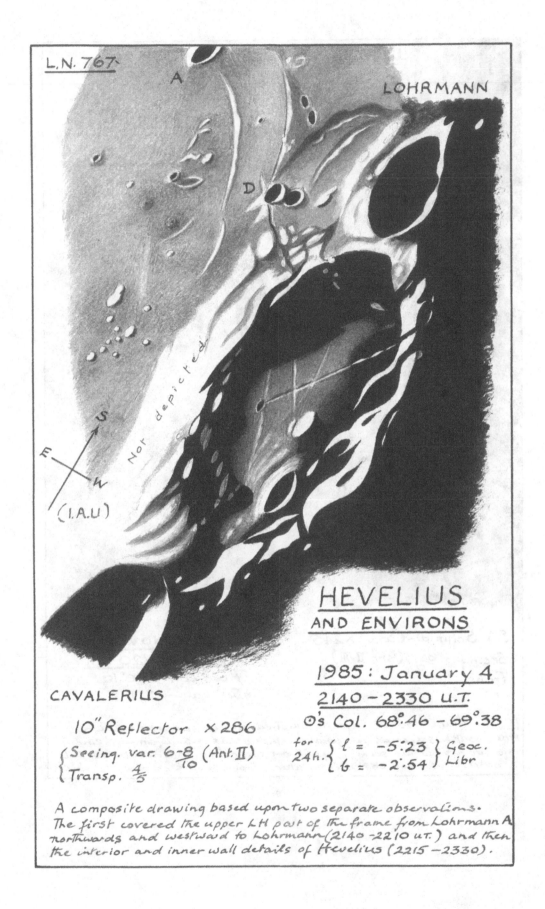

L.N. 767

A

LOHRMANN

D

Not depicted

S
E
W
(I.A.U)

CAVALERIUS

HEVELIUS
AND ENVIRONS

1985: January 4
2140 - 2330 U.T.
⊙'s Col. 68°.46 - 69°.38

10" Reflector ×286

{ Seeing. var. 6-8/10 (Ant. II)
{ Transp. 4/5

for
24h. { l = -5.23 } Geoc.
     { b = -2.54 } Libr

A composite drawing based upon two separate observations.
The first covered the upper L.H. part of the frame from Lohrmann A
northwards and westward to Lohrmann (2140 -2210 u.T.) and then
the interior and inner wall details of Hevelius (2215 -2330).

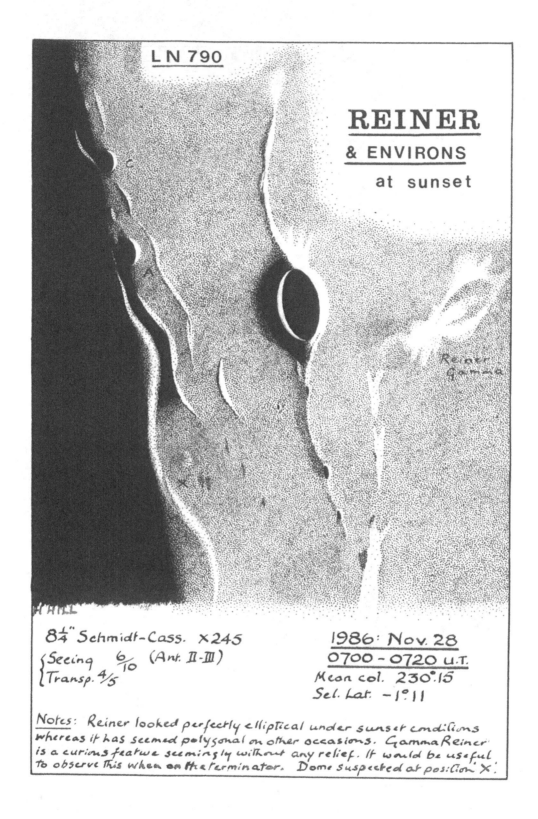

LN 790

REINER

& ENVIRONS

at sunset

Reiner Gamma

H HILL

$8\frac{1}{4}$" Schmidt-Cass. ×245

{Seeing $\frac{6}{10}$ (Ant. II-III)

{Transp. $\frac{4}{5}$

1986: Nov. 28

0700 - 0720 U.T.

Mean col. 230°.15

Sel. Lat. -1°.11

Notes: Reiner looked perfectly elliptical under sunset conditions whereas it has seemed polygonal on other occasions. Gamma Reiner is a curious feature seemingly without any relief. It would be useful to observe this when on the terminator. Dome suspected at position 'X'.

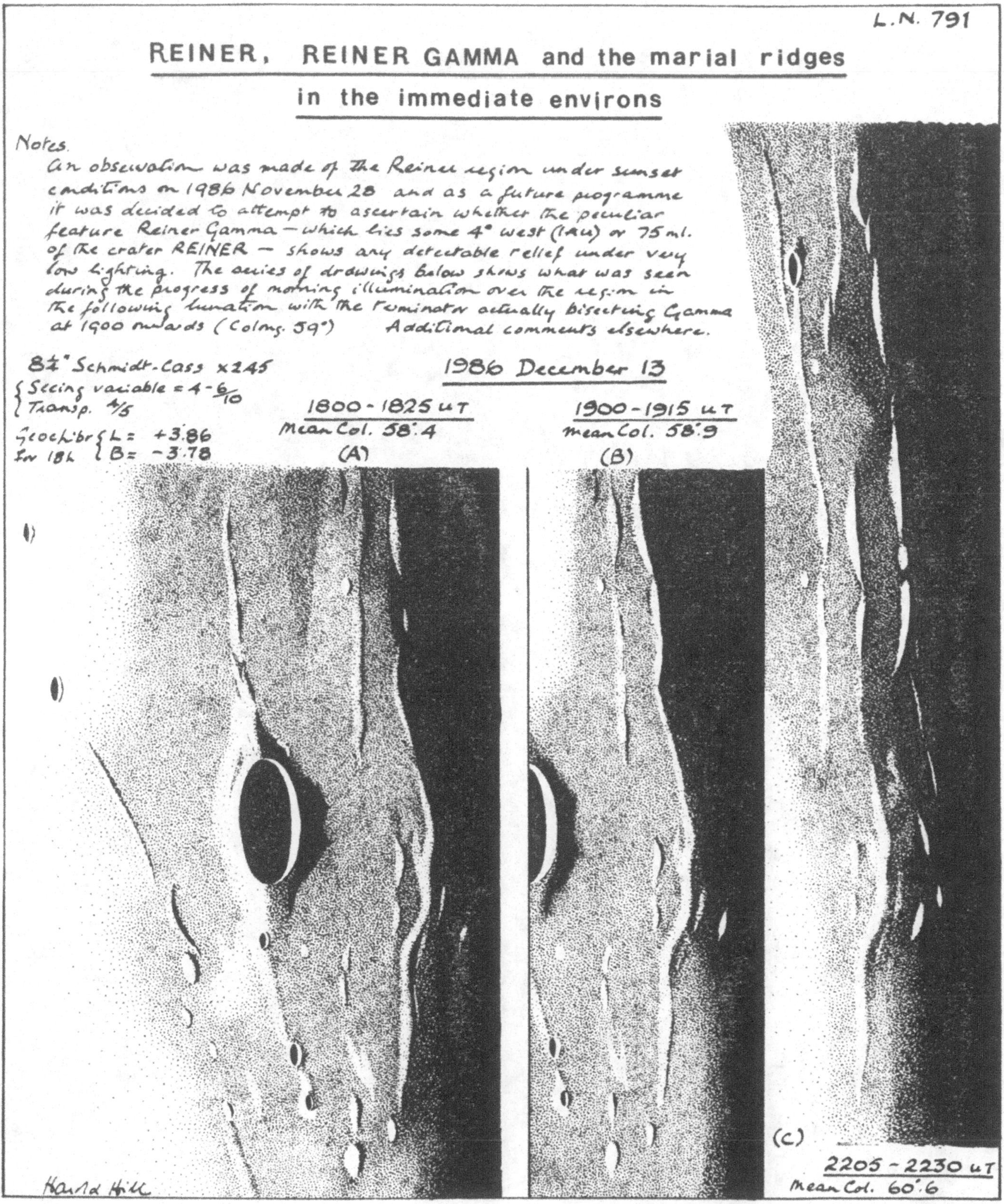

L.N. 791

REINER, REINER GAMMA and the marial ridges
in the immediate environs

Notes.

An observation was made of the Reiner region under sunset conditions on 1986 November 28 and as a future programme it was decided to attempt to ascertain whether the peculiar feature Reiner Gamma — which lies some 4° west (IAU) or 75 ml. of the crater REINER — shows any detectable relief under very low lighting. The series of drawings below shows what was seen during the progress of morning illumination over the region in the following lunation with the terminator actually bisecting Gamma at 1900 onwards (Colng. 59°)   Additional comments elsewhere.

8¼" Schmidt-Cass x245
{ Seeing variable = 4-6/10
{ Transp. 4/5

Geochibr { L = +3.86
For 18h { B = -3.78

1986 December 13

1800-1825 uT
Mean Col. 58°.4
(A)

1900-1915 uT
mean Col. 58°.9
(B)

2205 - 2230 uT
mean Col. 60°.6
(C)

Harold Hill

Quadrant II – Section 8

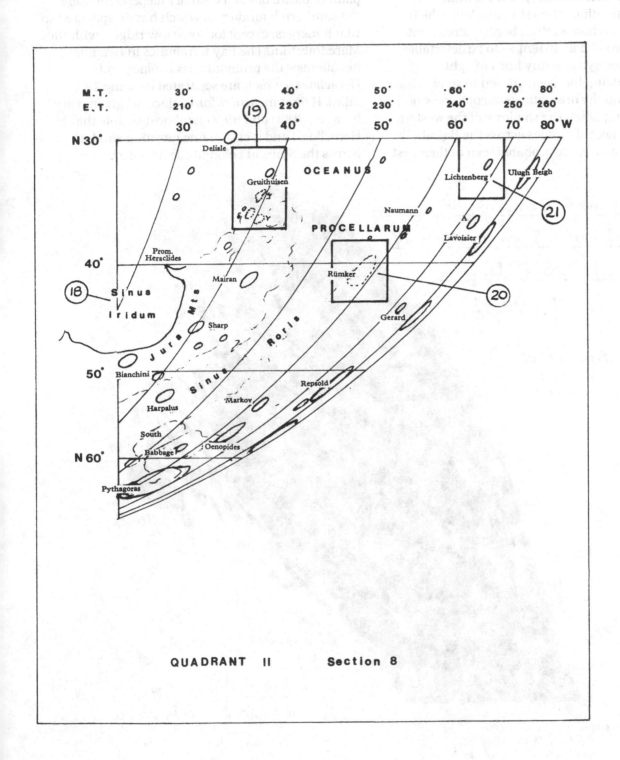

QUADRANT II   Section 8

Two sunset studies of the Sinus Iridum Region

Even the most casual lunar observer will be familiar with the magnificent Sinus Iridum – the 'Bay of Rainbows', as it is readily visible in the smallest of telescopes and a most striking feature under morning illumination when the Jura Mts, which form its northern and western border, stretch into the darkness to such an extent as to be detectable to a sharp naked eye as a tiny horn of light.

The first drawing has been chosen to show a less familiar, but equally dramatic, telescopic view of opposite lighting when the shadows of the western Jura cast their jagged profiles across the Bay at sunset. The shadows are probably seen at their best at the stage depicted here because shortly afterwards they engulf the Sinus completely.

The Sinus Iridum was originally a vast walled-plain or basin, or even a small independent Mare, the southern boundary of which has disappeared so that it merges, except for some low ridges, with the Mare Imbrium. The Bay terminates in two fine headlands – the promontories Laplace and Heraclides – which are separated by some 145 miles. Height measures for Laplace, as given, vary between 9000 and 9900 ft (almost double that of Heraclides) and it casts a fine morning shadow across the Sinus at colongitude 29°–30°.

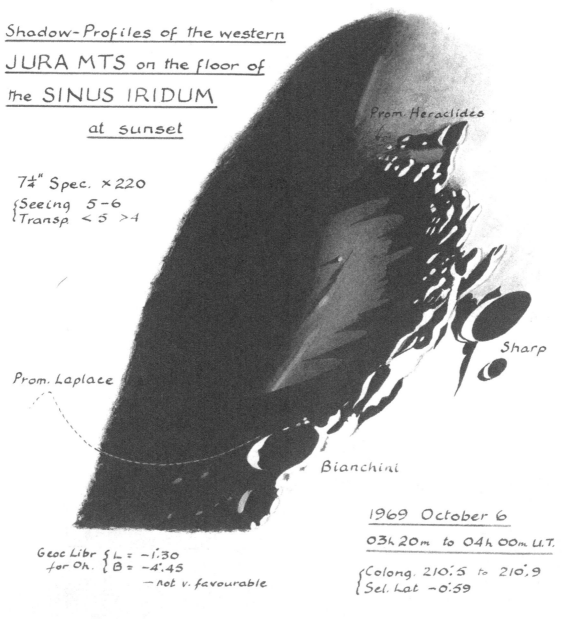

Shadow-Profiles of the western JURA MTS on the floor of the SINUS IRIDUM at sunset

7¼" Spec. ×220
{Seeing 5-6
{Transp. < 5 >4

Prom. Heraclides

Sharp

Prom. Laplace

Bianchini

1969 October 6
03h 20m to 04h 00m U.T.
{Colong. 210°.5 to 210°.9
{Sel. Lat. -0°.59

Geoc Libr {L = -1°.30
for 0h. {B = -4°.45
— not v. favourable

Dotted line indicates the approx. position of the unilluminated remainder of the "coastline".

The drawing made in 1972 shows sunset on the western portion of the Sinus with Prom. Laplace receiving the last rays of sunlight and the 28-mile crater Maupertuis to the north now almost completely shadow-filled. The low ridges and the craterlet Laplace A, 5½ miles in diameter, to the south are well seen at this stage but, though marking the southern limit of the Sinus, they do not follow closely the projected position of the ellipse if the original Bay had been truly circular, and it is questionable whether these represent remnants of a southern border.

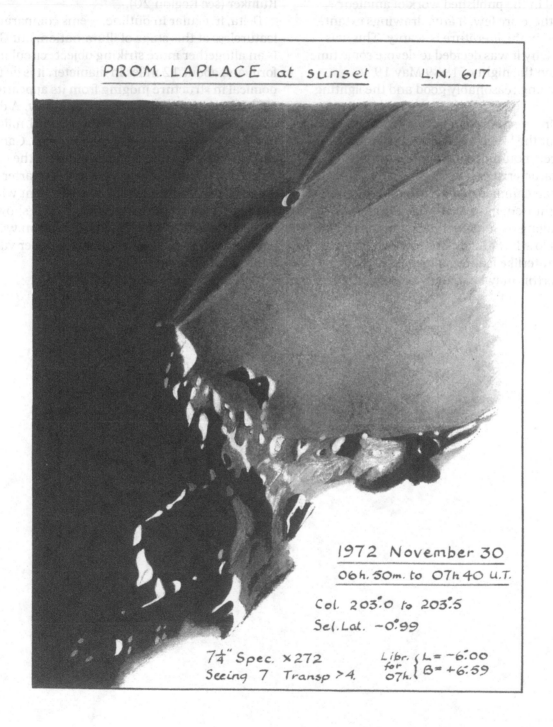

PROM. LAPLACE at sunset   L.N. 617

1972 November 30

06h. 50m. to 07h 40 U.T.

Col. 203°.0 to 203°.5

Sel. Lat. −0°.99

7¼" Spec. ×272
Seeing 7 Transp >4.

Libr. { L = −6°.00
for   { B = +6°.59
07h.

The Gruithuisen Group

One of the last observations made to be added to this portfolio. It is a region which does not appear to have figured in the published work of amateur lunarians; there are few, if any, drawings extant, and mention in the literature is scanty. This was one reason why it was decided to devote some time to the area on the night of 1989 May 19 – conditions being reasonably good and the lighting suitable.

A preliminary study such as this precludes anything but the briefest of comments except to say that the three mountains have interestingly different characteristics.

Zeta, situated immediately north of the $9\frac{1}{2}$ mile crater Gruithuisen, has at least three craterlets positioned along its southern edge and the block is scored with low N–S ridges, but there is some doubt about the craterlike feature at its northern extremity as this may be shadow contained between two spurs. Zeta gives the strong impression of being a small plateau not dissimilar in character to Rumker (see Region 20).

Delta, irregular in outline, seems comparatively featureless at this stage of illumination, but Gamma is an altogether more striking object; circular in form and about 12 miles in diameter, it is distinctly domical in structure judging from its appearance and the symmetrical shape of its shadow. A distinct summit craterlet, not much more than 1 mile in diameter, is situated just north of centre. Gamma at this early stage is much the brightest of the three mountains – particularly so in its NE quarter.

Inevitably, some differences are evident when the present record is compared with what is shown on available maps. It is likely that this region will repay examination in moderate apertures under varied lighting conditions.

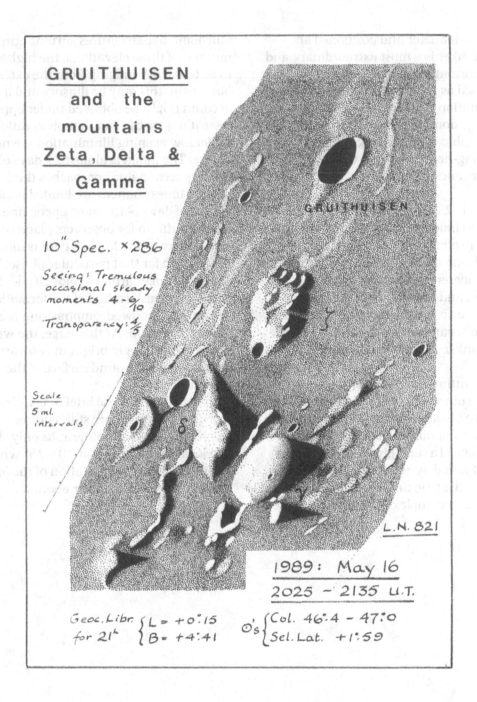

GRUITHUISEN
and the
mountains
Zeta, Delta &
Gamma

10" Spec. ×286

Seeing: Tremulous
occasional steady
moments 4-6/10

Transparency: 4/5

Scale
5 ml.
intervals

GRUITHUISEN

L.N. 821

1989: May 16
2025 - 2135 U.T.

Geoc. Libr. { L = +0°.15
for 21ʰ { B = +4°.41

☉'s { Col. 46°.4 - 47°.0
{ Sel. Lat. +1°.59

Rümker

About 34 miles in diameter and positioned at 41° N 58° W, Rumker is a most extraordinary and possibly unique formation which has been variously described as a 'complex of lunar domes', 'ruined ring formation', 'plateau', 'semi-ruined plateau' and 'large domical uplift'. It has also been likened, more graphically, to 'a large cindery mass' and 'a celestial slag-heap' – both of which aptly describe its appearance under certain conditions of oblique lighting.

Drawings nos. 1, 2 and 3 opposite were made on different dates and independently of each other. Even allowing for somewhat differing scales and conditions of lighting and libration, comparison shows puzzling differences but more especially in the interpretation and disposition of the smaller details. None of the observations was made under really favourable libratory circumstances, indeed, 1 and 1a show Rumker at almost maximum foreshortening.

Of the various authorities, Goodacre alone compares it to a 'ruined ring formation' which only fits its appearance at early sunrise (drawing). This is illusory because in a matter of hours the interior shadow quickly lifts (1a) and gives rise to a more lagoon-like effect noted by some observers. As the lighting proceeds, the true nature of the formation becomes evident as a complex of elevations, bosses,

and dome-like structures surmounting a low mound. Of these elevations, the highest would seem to be to the SW judging from the exterior shadow but, again, this may be illusory and a different account might be obtained under opposite lighting were it not for the fact that observations of Rumker under low evening illumination are not easy to secure. The Moon is almost 27 days old when the evening terminator approaches the formation and at such times chances are limited because of the advent of dawn – the most opportune time being during autumn for observers placed in the northern hemisphere as the Moon's path is most inclined to the horizon for that particular phase. The writer has caught Rumker around colongitude 225° when parts of the eastern half of the formation are becoming shadowed, emphasising the lagoon effect of the NE portion. At this stage, the westward-facing slopes are strikingly bright in contrast to the now darkening background surface of the Oceanus Procellarum.

In concluding these brief notes, observations of this unusual object are still required as the high resolution Orbiter photographs only show Rumker at a local solar altitude of 20°–25° which is too high for the effective interpretation of the low-relief features in which Rumker abounds.

RÜMKER
— some early morning aspects —

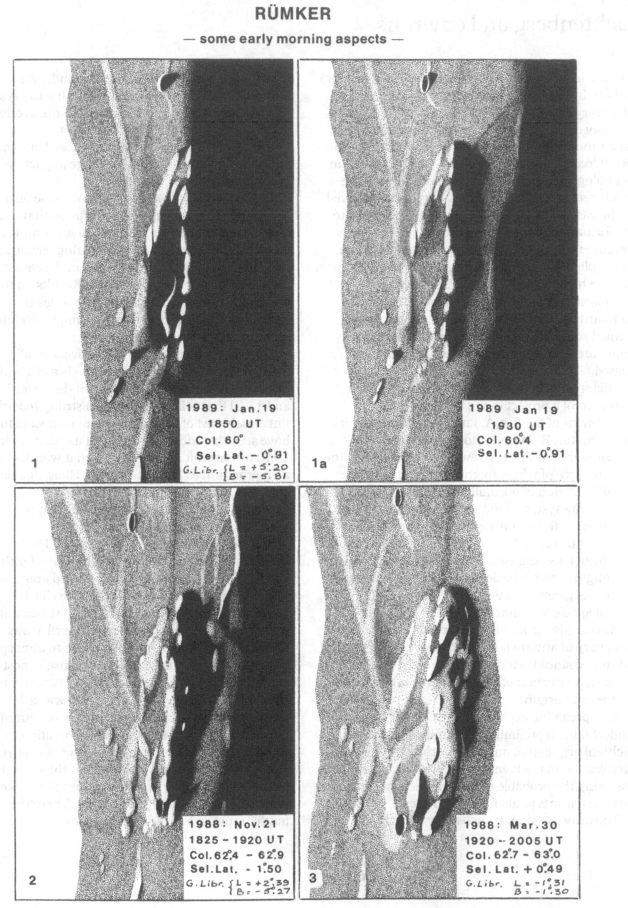

1989: Jan.19
1850 UT
Col. 60°
Sel. Lat. – 0°.91
G.Libr. {L = + 5°.20
{B = – 5°.81

1

1989 Jan 19
1930 UT
Col. 60°.4
Sel. Lat. – 0°.91

1a

1988: Nov. 21
1825 – 1920 UT
Col. 62°.4 – 62°.9
Sel. Lat. – 1°.50
G.Libr. {L = + 2°.39
{B = – 5°.27

2

1988: Mar. 30
1920 – 2005 UT
Col. 62°.7 – 63°.0
Sel. Lat. + 0°.49
G.Libr. L = – 1°.31
         B = – 1°.30

3

Lichtenberg and environs

An article in *Astronomy* Magazine for October 1983 written by P. H. Schultz of NASA's Lunar & Planetary Institute and P. D. Spudis of the Dept. of Geology of the Arizona State University contained an announcement that Lichtenberg is possibly the site of 'recent' lunar volcanism – recent, that is, in selenological terms. Briefly, the substance of the above report is that, since the lava sheets observed to the east and south of the crater are posterior to the formation of the Lichtenberg ray system because they obliterate the rays in that direction, the implication is that volcanism *followed* the impact event which caused the crater. As this is a comparatively 'young' crater by selenological standards . . . only 900 million years old, it was formed some 1000 million years *after* the period when lunar volcanism was considered to have ceased.

This startling conclusion led, in 1987, to the initiation of a project in the topographical department of the B.A.A. Lunar Section under its co-ordinator, R. Moseley, to search for:

(a) similar lava sheets elsewhere on the Moon, the presence of which might be indicated by the obliteration of normally continuous features, e.g. ray systems, and any significant interruption of uniformly toned areas of any extent, etc,

(b) 'fronts' associated with such lava flows which might possibly be detectable at appropriate times given good visual conditions and adequate instrumentation.

As regards the latter, the importance of telescopic discovery of any such fronts cannot be over-estimated since their disposition might give clues to the initial direction of the flow of the lava sheets and hence their origin.

This preamble explains the reason for the author undertaking a preliminary examination of the Lichtenberg region, namely, to study the general distribution of the lava sheet(s) with a view to assessing the probable selenographical position of any such fronts to aid future searches.

The drawing shown opposite was his first observation with this objective in mind, and although the lighting is too early for the ray system to make an appearance, the extent of the overlying dark lava field to the east and south is unmistakeable. Craterlets superimposed on the lava areas are obviously due to late stage impact events of a relatively minor character.

The sequence of five observations made on 1988 November 21 represent the attempt on that date to look for bright lines, arcuate or linear, coming into view with the advance of the morning terminator which might indicate the location of the eastern limit of the Lichtenberg lava flow. Results on this occasion were inconclusive but the series is included to suggest how the work might usefully be set out.

The Lichtenberg crater – a relatively small isolated ring 12 miles in diameter – is a celebrated object because it was here between the years 1830 and 1840 that Mädler observed a strong reddish tint closely east of the crater. No one appears to have seen this subsequently until its recovery in 1940 by Barcroft in the USA when it was described as 'a pronounced reddish-brown'. Haas, at a later date, observed the same effect and in 1951 Baum in the UK reported a rose-pink coloration which persisted for a time and then faded.

Between 0115 and 0320 on April 1 1988 (no significance in the date!) the author saw, for the first time in his experience, rosy-tinged areas fringing the northern edge of the lava sheet. The Moon had hitherto always looked quite neutral in colour to his eyes but the effect on April 1 was unmistakeable and certainly not due to atmospheric dispersion or other false effects, because a most careful check was made with bright objects, such as Lavoisier A. The telescope employed was a 10″ reflector using various eyepieces as a safeguard, but the colour persisted until weather conditions deteriorated. The area has been described as 'red-sensitive' but, if so, it is curious that the coloration is rarely in evidence, even when specially looked for under the most suitable conditions. There the matter must rest . . . . at least for now.

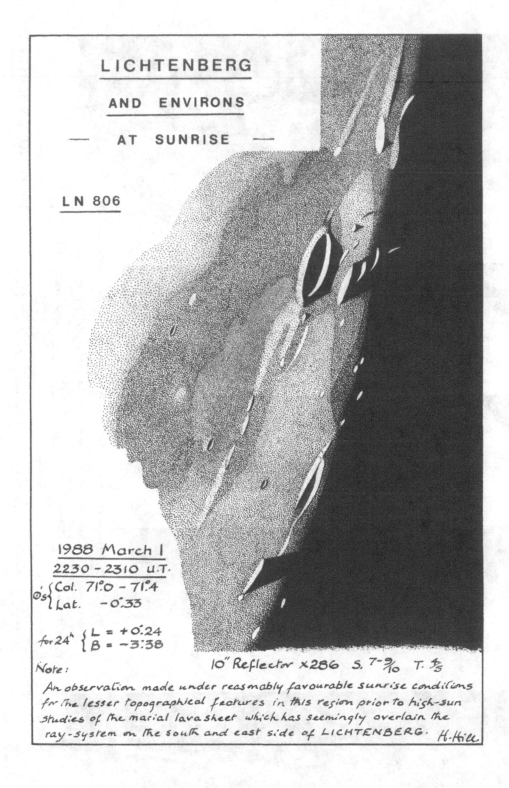

LICHTENBERG

AND ENVIRONS

— AT SUNRISE —

LN 806

1988 March 1
2230 - 2310 U.T.
⊙s { Col. 71°0 - 71°4
     { Lat. -0°33

for 24ʰ { L = +0°24
         { B = -3°38

Note:

10" Reflector ×286 S. 7-⁹⁄₁₀ T. ⁴⁄₅

An observation made under reasonably favourable sunrise conditions for the lesser topographical features in this region prior to high-sun studies of the marial lava sheet which has seemingly overlain the ray-system on the south and east side of LICHTENBERG. H. Hill

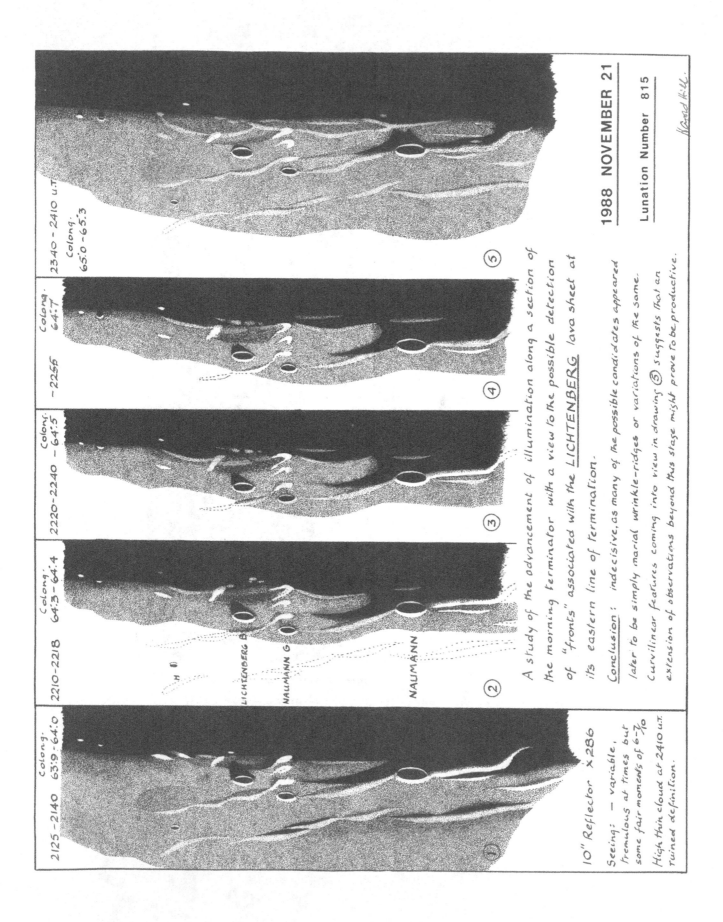

2125 – 2140   Colong. 63°.9 – 64°.0

2210 – 2218   Colong. 64°.3 – 64°.4

2220 – 2240 – Colong. 64°.5

– 2256   Colong. 64°.7

2340 – 2410 u.T.   Colong. 65°.0 – 65°.3

LICHTENBERG B
H
NAUMANN G
NAUMANN

① ② ③ ④ ⑤

1988 NOVEMBER 21

Lunation Number 815

A study of the advancement of illumination along a section of the morning terminator with a view to the possible detection of "fronts" associated with the *LICHTENBERG* lava sheet at its eastern line of termination.

Conclusion: indecisive, as many of the possible candidates appeared later to be simply marial wrinkle-ridges or variations of the same. Curvilinear features coming into view in drawing ⑤ suggests that an extension of observations beyond this stage might prove to be productive.

10" Reflector ×286

Seeing: – variable, tremulous at times but some fair moments of 6–7⁄10

High thin cloud at 2410 u.t. ruined definition.

96

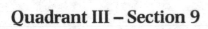

Quadrant III – Section 9

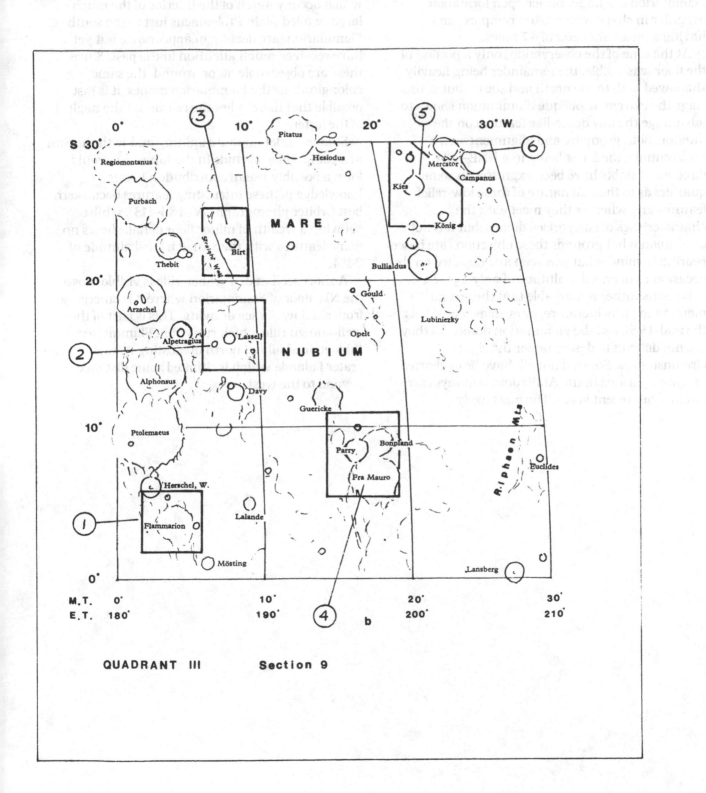

S 30°

0°   ③   10°   Pitatus   20°   ⑤   30° W

Regiomontanus   Hesiodus   Mercator   ⑥

Purbach   Kies   Campanus

MARE   König

Thebit   Birt   Straight Wall

20°   Bullialdus

Arzachel   Gould

②   Alpetragius   Lassell   Opelt   Lubiniezky

N U B I U M

Alphonsus   Davy

Guericke

10°   Ptolemaeus   Parry   Bonpland

Fra Mauro   Riphaen Mts

Herschel, W.   Euclides

①   Flammarion   Lalande

Mösting   Lansberg

0°

M.T.   0°   10°   20°   30°
E.T.   180°   190°   ④   200°   210°

b

QUADRANT III   Section 9

The Flammarion Domes

The coordinates of this walled-plain are 3° S 4° W and it is, therefore, placed quite near to the centre of the mean disc. Formerly known as Herschel F, Flammarion is a large, rather open formation, irregular in shape with broken ramparts and having a mean diameter of 47 miles.

At the time of the observation, only a portion of the floor was visible, the remainder being heavily shadowed both to the north and south, but at this stage the extremely oblique illumination shows to advantage the low dome-like features on the interior. Both grouping and alignment seems preferential – the latter being in a NNE–SSW direction. Doubts have been expressed in some quarters as to the real nature of these low-relief features and whether they meet with the characteristics of categorised domes, but it is not clear upon what grounds these objections are based bearing in mind what was seen and depicted on this occasion under a solar altitude of only $1\frac{1}{2}°-2°$.

It seems rather remarkable that there is no mention in early literature, or even as recently as the mid-1950s, of the Flammarion domes, as they are not difficult to descry under the right circumstances. Some, if not all, have been charted and incorporated in the ALPO dome surveys carried out in more recent years. The most likely explanation for their omission from earlier maps may lie in their essentially evanescent appearance. The numerous shallow saucer-shaped depressions which occupy much of the interior of the much larger walled-plain Ptolemaeus just to the south of Flammarion, are fleeting in appearance but yet have received much attention in the past. Since these are observable at, or around, the same colongitude as the Flammarion domes, it is just possible that therein lies the reason for the neglect of the latter.

Studies under opposite lighting and with the Sun at a comparable altitude in the lunar sky would form a possibly useful contribution to our knowledge of these interesting features because the best Orbiter photograph (IV–183–H3), whilst showing a wealth of minor floor detail, shows no dome features with the Sun at a local altitude of 24°.4.

A short section of a graben rille is visible crossing the NW floor of Flammarion where it is emerging from shadow . . . see drawing. This is part of the well-known rille which runs from Flammarion K (hidden in shadow) north of Mösting A, towards the crater Lalande which is situated some distance beyond to the west.

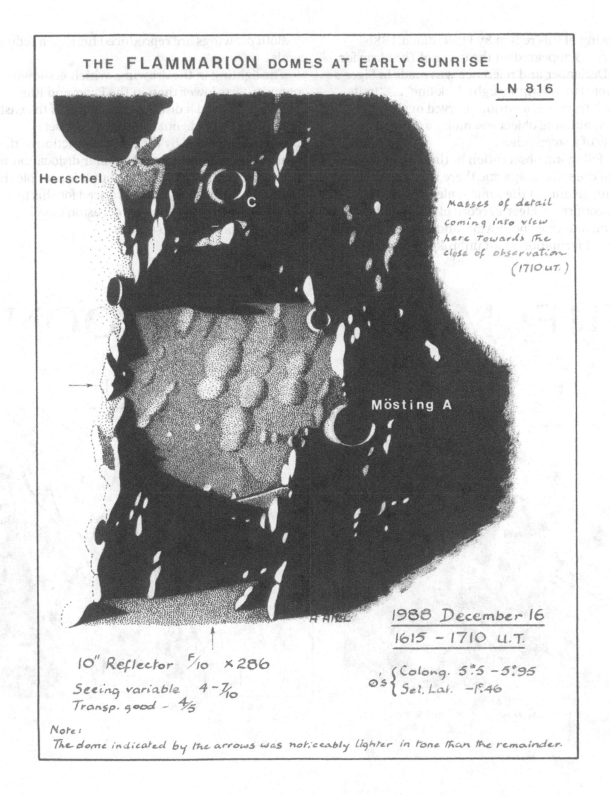

THE **FLAMMARION** DOMES AT EARLY SUNRISE

LN 816

Herschel

C

Masses of detail
coming into view
here towards the
close of observation
(1710 U.T.)

Mösting A

A HILL

1988 December 16
1615 – 1710 U.T.

10" Reflector ^{F}/10 × 286

Seeing variable 4 – 7/10
Transp. good – 4/5

O's { Colong. 5°.5 – 5°.95
{ Sel. Lat. –1°.46

Note:
The dome indicated by the arrows was noticeably lighter in tone than the remainder.

Lassell and environs

A drawing of this region by Elger, dated 1886 January 13 appeared on the cover of *The Moon* for 1956 December and reference was made in Elger's own notes to 'a long straight dark line' . . . 'In its course it traverses a curious curved oval-shaped ring'. Again, 'the object resembles a fault, or shadow of a steep ridge'.

In a follow-up observation in the 1957 October issue of the same magazine there was a cover drawing on almost the same scale by W. L. Rae which confirmed Elger's record of this feature, showing it to cross both walls and interior of the elongated formation in an unbroken manner.

Both drawings are reproduced here on a reduced scale.

The lighting in the drawing, which is shown opposite, was lower than in the Elger and Rae observations and a distinct dislocation of the east wall could be made out (see arrowed inset) presumably formed by the slumping action of the fault. Though looked for, no similar dislocation was detectable on the western side but it is possible that lighting was insufficiently advanced for this to show. Unfortunately, on this occasion cloud brought proceedings to a close.

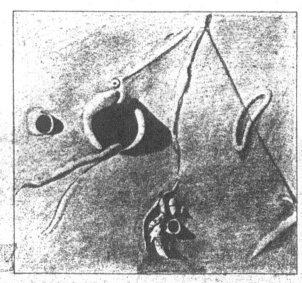

THE MOON

VOLUME 5. No 2.

DECEMBER 1956.

LASSELL
1886 JANUARY 15.   8½" Refl. x 340.
T. G. Elger.

THE MOON

VOL 6. No 1

OCTOBER 1957.

LASSELL AND SURROUNDING AREA
9¼-inch Refl.   W. L. Rae.

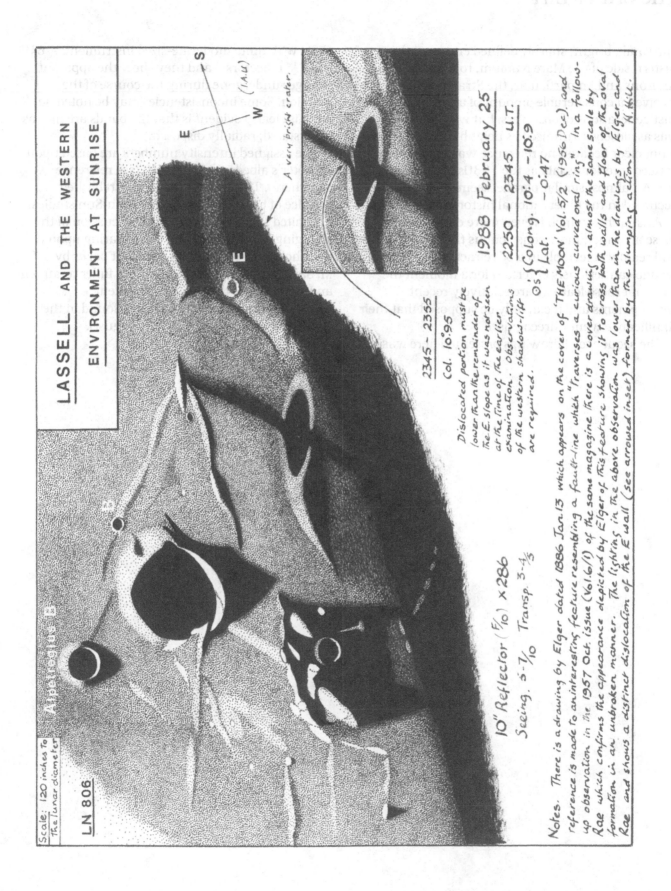

Scale: 120 inches to
the lunar diameter

LN 806

Abenezra B

LASSELL AND THE WESTERN
ENVIRONMENT AT SUNRISE

S
W (I.A.U.)
E

A very bright crater.

E

B

2345 - 2355
Col. 16°.95

Dislocated portion must be
lower than the remainder of
the E. slope as it was not seen
at the time of the earlier
examination. Observations
of the western shadow-lift
are required.

1988 February 25

2250 - 2345 U.T.

2345 U.T.
Ⓞ's { Colong. 10°.4 -10°.9
       Lat.   -0°.47

10" Reflector (⁶/₁₀) ×286
Seeing. 5-7/10 Transp. 3-4/5

Notes. There is a drawing by Elger dated 1886 Jan.13 which appears on the cover of 'THE MOON' Vol.5/2 (1956 Dec.) and
reference is made to an interesting feature resembling a fault-line which "traverses a curious curved oval ring". In a follow-
up observation in the 1957 Oct. issue (Vol.6/1) of the same magazine there is a cover drawing on almost the same scale by
Rae which confirms the appearance depicted by Elger of this feature showing it to cross both walls and floor of the oval
formation in an unbroken manner. The lighting in the above observation was lower than in the drawings by Elger and
Rae and shows a distinct dislocation of the E. wall (see arrowed inset) formed by the slumping action. K.Hill.

The crater Birt

The small, bright, sharply defined crater on the eastern side of the Mare Nubium, together with its environs and, in particular, the Straight Wall, received a considerable amount of attention from past generations of observers, but whether Birt itself was as closely scrutinised as neighbouring features is not clear because no reference was made to its banded interior as late as the 1930s of the Goodacre era. An almost similar parallel seems to have occurred in the observational history of the bands in Aristarchus which is even more curious since these are much more conspicuous than those in Birt and readily detectable in a 3″ refractor. However, no one can seriously entertain for a moment the idea that these features are relatively recent developments and it can only be supposed that their significance went unrecognised.

The sequence of drawings presented here was made with quite moderate-sized instruments – $6\frac{1}{2}$″ and $7\frac{1}{2}$″ reflectors – and they show the apparent changes undergone during the course of the lunation. Some inconsistencies may be noted but what is clearly evident is that the bands are not, as often stated, radially disposed.

The assigned intensity numbers are based upon Schröter's albedo scale of 0–10 (see reference elsewhere). This follows the long-established practice of US observers in their high-Sun studies.

Limited space precludes a full discussion of the changing appearances and their bearing upon the morphology of Birt, but an excellent paper by M. Savill & R. McKay which gives a fuller treatment and summarises the results of amateur investigations of the crater may be found in the *B.A.A. Journal*, **91** No. 5 (1981 August).

BIRT

— changing aspects during the lunar day —

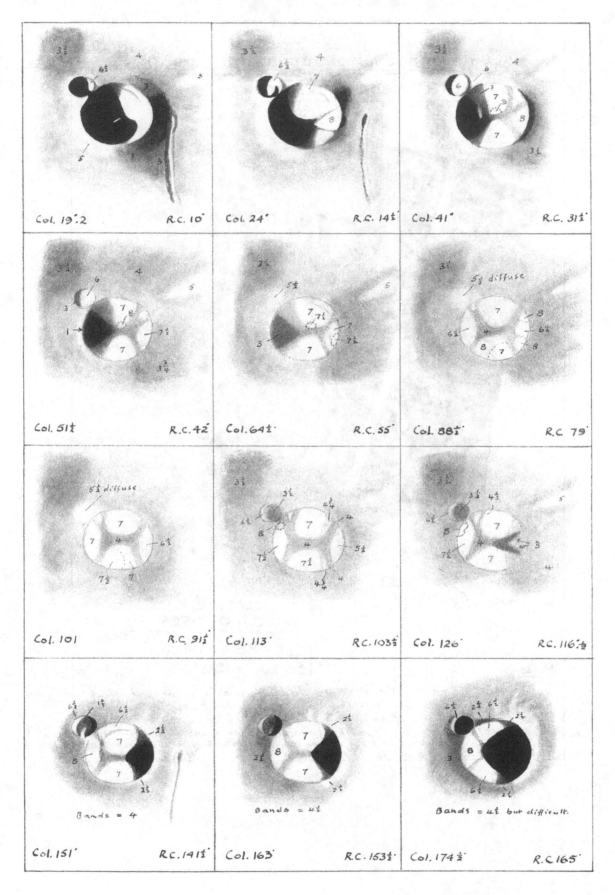

Late stage of illumination on
BIRT and environs —

L.N. 802

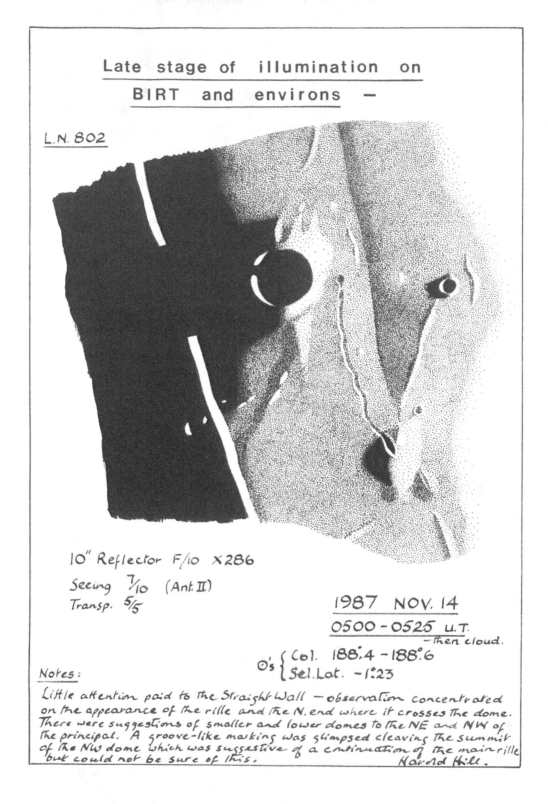

10" Reflector F/10 X286

Seeing $^7/_{10}$ (Ant. II)

Transp. $^5/_5$

1987 NOV. 14

0500 - 0525 U.T.

— then cloud.

⊙'s $\begin{cases} Col. & 188°.4 - 188°.6 \\ Sel. Lat. & -1°.23 \end{cases}$

Notes:

Little attention paid to the Straight Wall — observation concentrated on the appearance of the rille and the N. end where it crosses the dome. There were suggestions of smaller and lower domes to the NE and NW of the principal. A groove-like marking was glimpsed cleaving the summit of the NW dome which was suggestive of a continuation of the main rille but could not be sure of this.

Harold Hill.

The Fra Mauro Group

The two drawings shown are from a series of observations of the Fra Mauro group, the mean position of which is situated at 7° S 17°W.

That of Parry with approaching evening terminator had to be worked very quickly and the shadow lengths are depicted as at the commencement of observation. Less than half-an-hour later, the area immediately east of Parry was in almost complete darkness and the shadow profiles of the western rampart had encroached very considerably on the floor.

With regard to the observation of 1988 April 25 made under opposite illumination, the notes written at the telescope read:

'Sharply demarcated areas of different tones, particularly on the interior of Parry, NE of Parry A, NE of the ''horse-shoe'' (M), and the surface SE of Parry F. The scalloped or serrated northern edge of Parry's N border is difficult to define'.

Further notes made later after completion read:

'An instance where, despite indifferent definition at the beginning, it was worthwhile proceeding with the general outline of the region for eventually there was a steady improvement in the seeing and, later on when it became excellent (9), the framework was already drawn for the insertion of the more delicate detail in its correct position.'

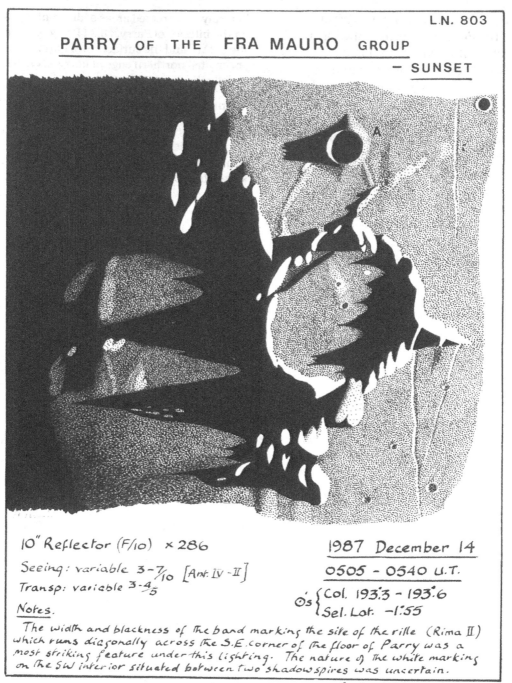

L.N. 803

PARRY OF THE FRA MAURO GROUP
— SUNSET

10" Reflector (F/10) ×286

Seeing: variable 3-7/10 [Ant. IV - II]

Transp: variable 3-4⅕

Notes.

1987 December 14

0505 - 0540 U.T.

☉'s { Col. 193°·3 - 193°·6

Sel. Lat. −1°·55

The width and blackness of the band marking the site of the rille (Rima II) which runs diagonally across the S.E. corner of the floor of Parry was a most striking feature under this lighting. The nature of the white marking on the SW interior situated between two shadow spires was uncertain.

Scale: 118 inches to lunar dia.

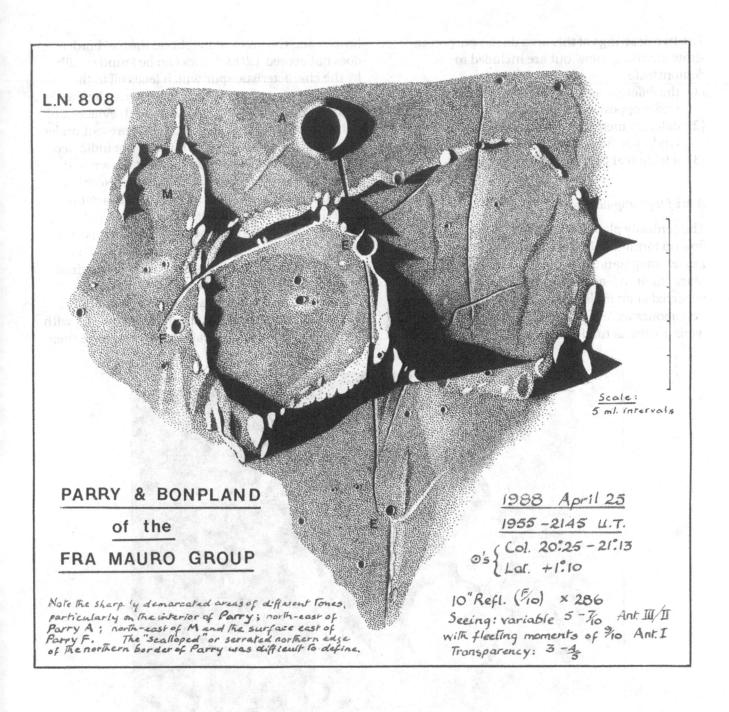

L.N. 808

PARRY & BONPLAND
of the
FRA MAURO GROUP

Note the sharp.'ly demarcated areas of different tones,
particularly on the interior of Parry; north-east of
Parry A; north-east of M and the surface east of
Parry F.   The "scalloped" or serrated northern edge
of the northern border of Parry was difficult to define.

Scale:
5 ml. intervals

1988 April 25
1955 -2145 U.T.
⊙'s { Col. 20°25 - 21°13
        Lat. +1°10

10"Refl. (F/10) × 286
Seeing: variable 5 -7/10   Ant. III/II
with fleeting moments of 9/10 Ant. I
Transparency: 3 -4/5

The Kies, Kies A and König Region

The two drawings of this area do not purport to show anything 'new' but are included to demonstrate:

(1) the different aspect of the same general region under opposite illumination;
(2) different methods of portraying the lunar landscape when making 'finished' sketches;
(3) a technical point regarding lighting.

A brief topographical note

The centrally placed Kies, 28 miles in diameter is a flooded formation by reason of the intrusion of molten magmatic material from the surrounding Mare Nubium – an event which must have occurred at an earlier period than when its lesser neighbours were formed. Where sections of the walls are not actually missing, they are low and broken and the greatest height on the SW border does not exceed 1200 ft. Kies can be found readily by the characteristic spur which leads off to the south for a short distance and which may be the remains of an adjoining ring. Although evidence of a similar rudimentary spur or ridge is present under evening light on the N border, this is not indicated on the morning observation, nor is it shown with certainty on any of the available charts. Note the well-known and well-formed dome with summit craterlet or pit WSW of Kies.

Notes will be found elsewhere in the first part of this portfolio dealing with the preparation of lunar drawings and the various stages from the original made at the telescope through to the 'finished' product. The Kies sketches show two techniques: the one employs stippling, the other Indian ink with dilutions of same for the half-tones. Both have their

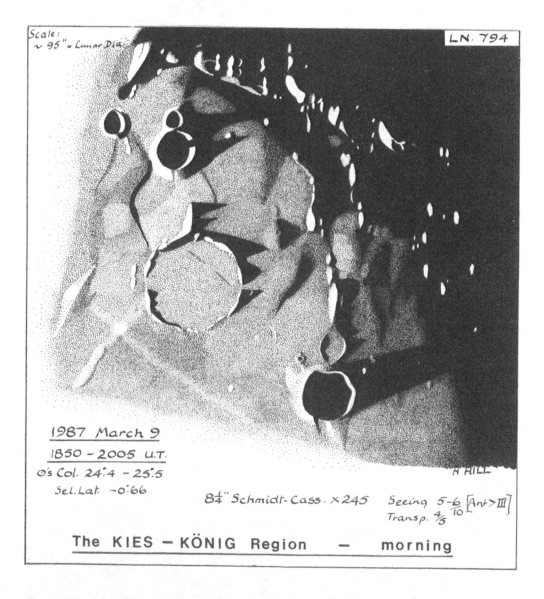

Scale: ∼ 95" = Lunar Dia.

LN. 794

1987 March 9
1850 – 2005 U.T.
O's Col. 24°.4 – 25°.5
Sel. Lat. ∼0°.66

8¼" Schmidt-Cass. × 245

Seeing 5-6 [Ant > III]
Transp. 4/5

H HILL

The KIES — KÖNIG Region   —   morning

own particular advantages, but the author prefers the latter as being more truly representative of the lunar surface as seen in the telescope.

The two observations exemplify the fact that, except in the immediate equatorial zones of the Moon, the value of the Sun's colongitude is not an exact measure of phase. In both instances, the distance of Kies from the theoretical terminator as given by the derived colongitude values is very nearly the same, but it will be seen that there is considerable difference in shadow length thrown by the crater walls and this is not due solely to variable height. Although Kies lies only some 26° south of the lunar equator it is, nevertheless, subject to some degree to the seasonal variations in the solar

altitude caused by a $1\frac{1}{2}$° tilt in the Moon's axis which is analogous to that of the Earth ($23\frac{1}{2}$°).

Shadows are longer in the 1983 drawing – the Sun being in the lunar northern hemisphere at lat. +1°.12 whereas, at the time the 1987 observation was made, the Sun was in the opposite hemisphere at −0°.66 . . . a material difference in local altitude of the order of 1°.8.

It will be appreciated that, in still higher selenographical latitudes, the effect is greater and increasingly so towards the poles. This has to be taken into account when confronted with what, at first sight, may appear to be observational inconsistencies when results at a given colongitude are compared.

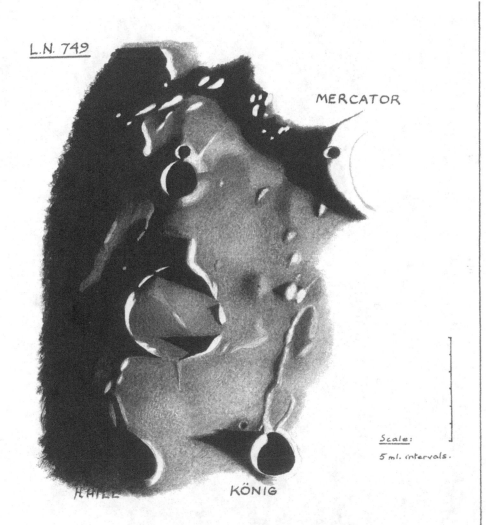

L.N. 749

MERCATOR

Scale:
5 ml. intervals.

KHIEL   KÖNIG

THE KIES & KIES A REGION

$8\frac{1}{4}$" Schmidt-Cass. ×245

Seeing $\frac{5}{10}$ Transp. $\frac{3}{5}$

1983 AUG. 3
0245 - 0330 U.T.
{ Col. 200°.4 - 200°.7
{ Sel. Lat. +1°. 12

Mercator and Campanus

Excellent accounts of the topographical features associated with these craters are to be found in the descriptive guides to the Moon by Goodacre and Wilkins & Moore.

Some lunar scenes, when caught at their most graphic, almost compel representation but very often they give limited information, as in the present case, due to extensive shadow.

Mercator and Campanus are a good example of the not infrequent occurrence of a close pair of ringed formations very similar in size (29 miles in diameter in this instance) and having other characteristics in common. Other pairings which are either in contact or close proximity to each other are Sabine & Ritter, Leverrier & Helicon, Rost & Rost A, Casatus & Klaproth, Billy & Hansteen, Azophi & Abenezra, Rosenberger & Vlacq, Aliacensis & Vlacq, Metius & Fabricius, Steinheil & Watt, with smaller examples such as the Messiers, Beer & Feuillée whilst the number of still smaller pairings is legion.

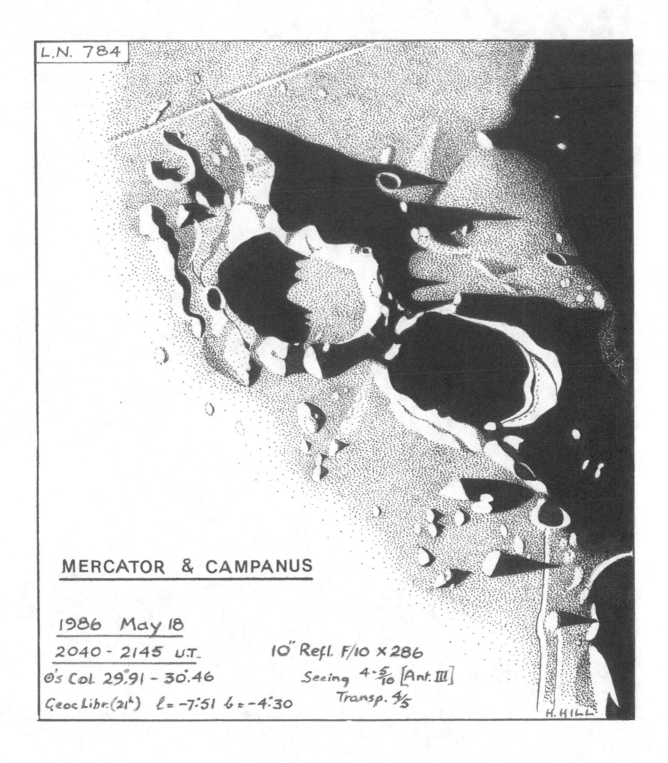

L.N. 784

MERCATOR & CAMPANUS

1986 May 18

2040 - 2145 U.T.   10" Refl. F/10 × 286

⊙'s Col. 29°.91 - 30°.46   Seeing 4·$\frac{5}{10}$ [Ant. III]

Geoc.Libr.(21ʰ) l = -7°51 b = -4°30   Transp. 4/5

H. HILL

Quadrant III – Section 10

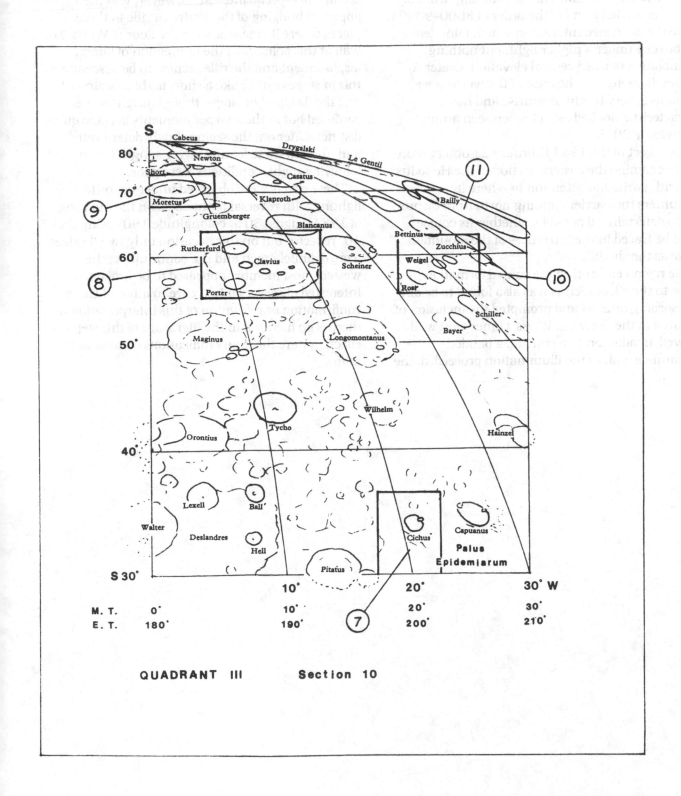

S

80°
70°
60°
50°
40°
S 30°

Cabeus
Drygalski
Le Gentil
Newton
Short
Casatus
Moretus
Klaproth
Gruemberger
Blancanus
Rutherfurd
Clavius
Scheiner
Porter
Maginus
Longomontanus
Wilhelm
Tycho
Orontius
Lexell
Ball'
Walter
Deslandres
Hell
Pitatus
Cichus
Capuanus
Palus
Epidemiarum
Bailly
Bettinus
Zucchius
Weigel
Rost
Schiller
Bayer
Hainzel

⑨ ⑧ ⑪ ⑩ ⑦

10°   20°   30° W

M. T.   0°   10°   20°   30°
E. T.   180°   190°   200°   210°

QUADRANT III   Section 10

The Cichus – Weiss Region

Cichus is a well-defined crater some 25 miles in diameter situated at the southern end of the mountain barrier dividing the Palus Epidemiarum from the Mare Nubium. The surrounding walls of the crater are lofty and of the order of 8000–9000 ft above the depressed interior on which faint details can be seen under a higher light, but nothing resembling a marked central elevation. Crater A, 7 miles diameter, on the crest of the western wall of Cichus is very bright at sunrise and has a characteristic hooked aspect when seen around colongitude 20°.5.

The object of the 1983 February 21 observation was to examine the eastern section of the Hesiodus rille with particular attention to where it encounters the barrier running north from Cichus and to determine, if possible, whether its course could be traced in the interstices of the mountain range as the shadows lifted.

The region embracing the ruined formation Weiss to the NE of Cichus was also found to be one of absorbing interest and prompted the inclusion of this area in the drawing. Weiss's inner west wall showed, in addition to terracing, a banded appearance and, as the illumination proceeded, the narrow defile of a rille which runs approximately parallel to the larger Hesiodus rille, could be discerned by its penetration of the barrier. What became of especial interest, however, was the apparent bridging of the southern rille in three places where it emerges on to the floor of Weiss. The wall at this point gave the impression of lateral displacement and the rille seemed to be descending this in stages rather like a short flight of stairs. At first the 'bridges' or 'steps' though bright, were confused but in the sharper moments became quite distinct. Later on, the seeing slowly deteriorated and, though these features were sought for on the following night, conditions were poor.

These effects were observed under opposite lighting with reversal of shadows on the morning of 1984 August 30 at colongitude 190° using the 10″ reflector but on this occasion only two 'bridges' were definitely seen and the continuity of the westernmost member appeared to be broken. Interested observers might care to attempt either confirmation or refutation of this interpretation as there is no mention in the literature of this stepped effect, nor are there any indications on available maps.

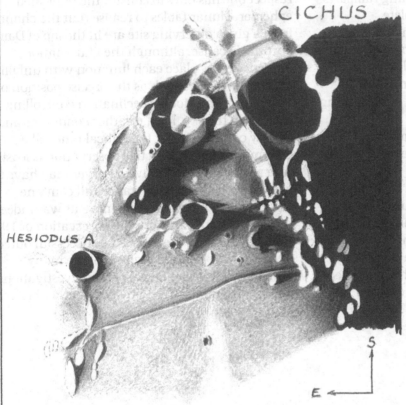

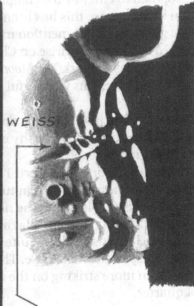

LN 744

CICHUS

WEISS

HESIODUS A

S

E

Bridging of the rille
north of CICHUS where
it cuts through the
W border of WEISS. As
seen at 2320 u.t. 5⁷/₁₀
(Col. 21°7)

The eastern portion of the HESIODUS RILLE

under morning illumination.

1983 Feb. 21
2120 - 2230 u.t.
2335

8¼" Schmidt-Cass. ×245

Seeing: variable 5 - 8/10 Transp. 4/5

Colongitude limits:
20°7 - 21°3

Clavius

This vast formation needs no introduction to even the most casual telescopist as it is one of the finest lunar objects when viewed under low illumination, not only by reason of its size (some 145 miles across) but also because of the wealth of varied detail on its convex floor which is depressed 16 000 ft below the massive surrounding ramparts. It is not proposed to enter into a complete description here because this has been ably dealt with in the guidebooks but mention must be made of the spirited account of sunrise on Clavius given by Neison in his classic work *The Moon* as this epitomises to the letter this wonderful and engaging scene.

No drawing can do justice to this unfolding drama nor, indeed, can photography except perhaps at the highest level.

The drawing made on 1973 April 11 was an effort, albeit a frustrating one, to capture the essentials of the visual picture at the time when the great bay of darkness created by Clavius at the terminator is so large as to blunt quite perceptibly the southern cusp to the naked eye. The same effect is, perhaps, even more striking on the crescent just after last quarter.

The author's own special interest in Clavius has been directed towards a study of the great east wall under sunset conditions when it is brilliantly lit and also to capture those rather infrequent opportunities of observing last light along the very crest itself – a challenging assignment. In this respect one has only to consult the published ephemeral lunar tables to realise that the chances from a given observing site are in the lap of Dame Fortune because, although the illumination sequence takes place each lunation with unfailing regularity, factors such as the precise position of the terminator, the Moon's declination controlling its altitude in the sky, local weather conditions and quality of the seeing at the critical time, all determine whether effective observation is feasible. As a result, the prospective observer may have to wait *years* before circumstances all combine favourably to meet requirements – as was indeed the case with the author. The observation of 1988 November 3 was, therefore, the successful outcome of a long wait – entirely typical of the many problems which can beset specific investigations, not only in lunar astronomy but in other branches of the science.

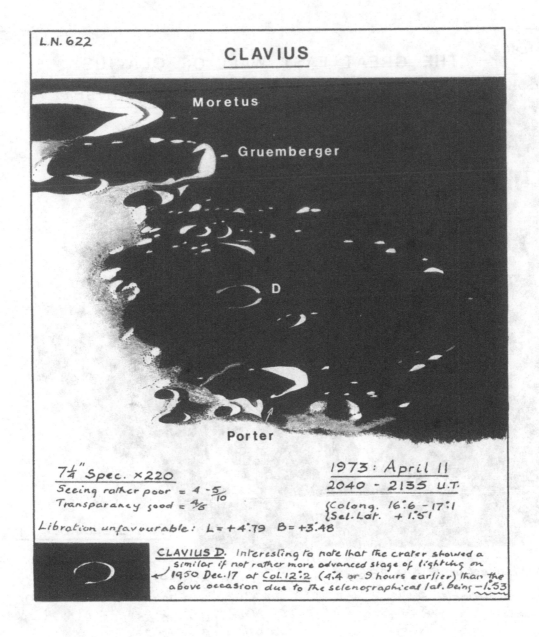

L.N. 622

CLAVIUS

Moretus

Gruemberger

D

Porter

$7\frac{1}{4}"$ Spec. ×220

Seeing rather poor = $4-\frac{5}{10}$

Transparancy good = $\frac{4}{5}$

Libration unfavourable: L = +4°79 B = +3°48

1973: April 11

2040 - 2135 U.T.

{Colong. 16°6 - 17°1

{Sel. Lat. + 1°51

CLAVIUS D. Interesting to note that the crater showed a similar if not rather more advanced stage of lighting on 1950 Dec. 17 at Col. 12°2 (4°4 or 9 hours earlier) than the above occasion due to the selenographical lat. being -1°53

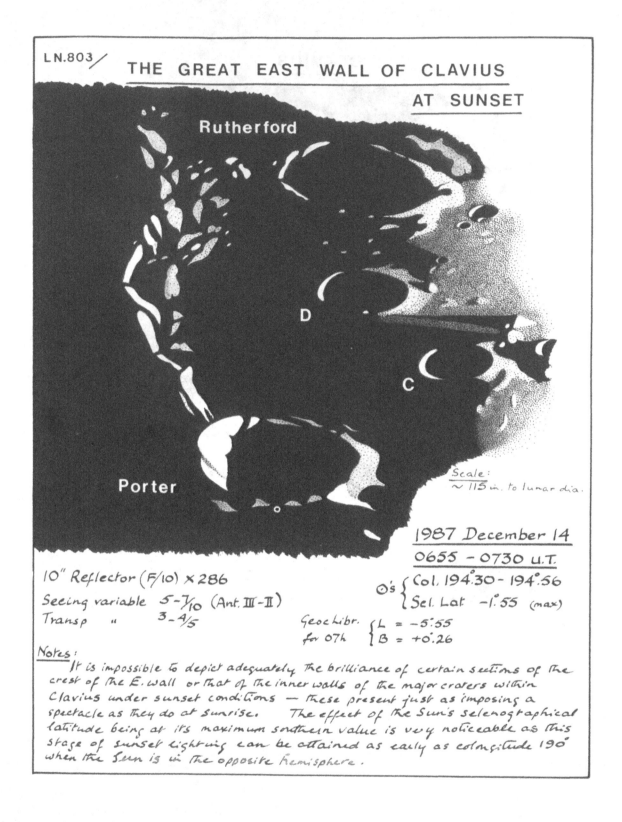

LN.803

THE GREAT EAST WALL OF CLAVIUS
AT SUNSET

Rutherford

D

C

Porter

Scale:
~ 115 in. to lunar dia.

1987 December 14
0655 - 0730 U.T.

⊙'s { Col. 194°.30 - 194°.56
      Sel. Lat. -1°.55 (max)

10" Reflector (F/10) × 286

Seeing variable 5 - 7/10 (Ant. III - II)
Transp   "   3 - 4/5

Geoc. Libr. { L = -5°.55
for 07h   { B = +0°.26

Notes:
   It is impossible to depict adequately the brilliance of certain sections of the
crest of the E. wall or that of the inner walls of the major craters within
Clavius under sunset conditions — these present just as imposing a
spectacle as they do at sunrise.   The effect of the Sun's selenographical
latitude being at it's maximum southern value is very noticeable as this
stage of sunset lighting can be attained as early as colongitude 190°
when the Sun is in the opposite hemisphere.

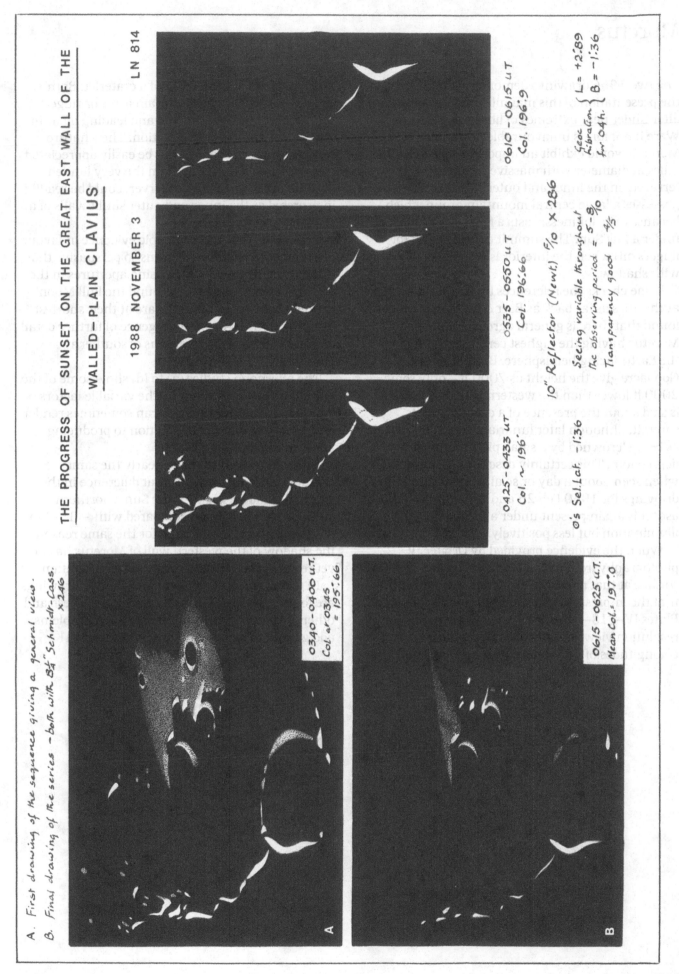

THE PROGRESS OF SUNSET ON THE GREAT EAST WALL OF THE
WALLED—PLAIN CLAVIUS

1988 NOVEMBER 3

LN 814

0610 - 0615 u.T
Col. 196.9

0535 - 0550 u.T
Col. 196.66

0425 - 0433 u.T
Col. ∼ 196°

0340 - 0400 u.T.
Col. at 0345 = 195°.66

0615 - 0625 u.T.
Mean Col. = 197°.0

10" Reflector (Newt.) f/10 × 286

Seeing variable throughout
the observing period = 5-8%
Transparency good = 4/5

⊙'s Sel.Lat. -1.36

Geoc. Libration { L = +2°.89 B = -1°.36 } for 04h.

A. First drawing of the sequence giving a general view.
B. Final drawing of the series - both with 8¼" Schmidt-Cass. × 246

A

B

Moretus

The two 1966 drawings demonstrate how greatly the presentation of this magnificent formation can alter under near extremes of libration in latitude. Were it not for its unfavourable position at 70° S, Moretus would exhibit an imposing spectacle: 75 miles in diameter with massive ramparts, deeply terraced on the inner and outer slopes, it also possesses a large central mountain group which, because of its loftiness, casts a fine spire of shadow under a low Sun. The summit appears early (or lingers on) when the interior is completely filled with shadow.

If one checks the references by various lunar authorities as far back as Beer & Mädler, it will be found that there is general agreement about Moretus having the highest central mountain on the Earth-facing hemisphere. Both Elger and Goodacre give the height as 7000 ft – only some 2000 ft lower than the western wall, but nothing is said about the presence of a craterlet on its summit, although later lunarians have described it as being 'crowned by a small pit' or 'a shallow depression'. This certainly describes its appearance when seen about a day or so after local sunrise (see drawings for 1950 Feb.26 and 27). The crateriform aspect is again present under afternoon illumination but less positively.

When the evidence provided by Orbiter IV photography from a near-vertical position over the surface at that point is considered, it would appear that the impression of a summit crater is illusory. Plates IV–118–H2 and 130–H2 show high-resolution views at solar selenographical colongitudes of 36° and 48° respectively, and neither gives any indication of a craterlet. Both show what seem to be mountain arms or ridge spurs emerging from shadow and leading off from the central mass in a SE direction. Their nature seems indisputable and it can be easily appreciated that these, seen obliquely from the very limited position of the terrestrial observer, could be readily interpreted as the inner and outer sunlit walls of a small ring.

Yet despite this, the telescopic evidence in favour of a crateriform feature remains so persuasive that it calls for a renewed attack using apertures in the 12″–18″ range, especially around the Full Moon period when shadow lengths are at their shortest resulting in the possible emergence of further detail at times of optimum conditions of southern libration.

The afternoon studies (a) to (d) show some of the odd effects brought about by the variable factors of lighting and libration which can sometimes render interpretation difficult in addition to producing apparent inconsistencies.

Sketches (c) and (d) have nearly the same colongitude value but the great difference in the length of shadow is due to the Sun's northern latitiude ($+1°01$) in (c) compared with $-1°33$ in (d). It will also be noted that, for the same reason, the shadow of the western wall of Moretus has not yet reached the vicinity of the central mountain.

In a long series of observations, no two drawings are found to be precisely alike because of these and other physical circumstances, but such problems add greatly to the challenge of investigation!

L.N. 543

MORETUS

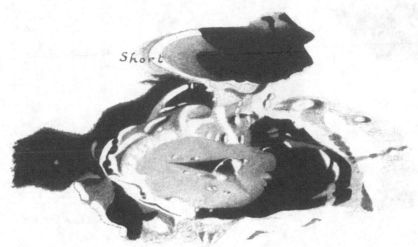

Short

7¼" Spec. ×292

Geoc   {L = −2°.64
Libr.   {B = −6°.58
(04ʰ)

{Seeing. 5 – 6
{Transp. < 4

1966: December 4

03ʰ 15ᵐ to 05ʰ 00ᵐ U.T.

Colong. 167°.8 to 168°.7

L.N. 544

1966: December 20

18.50 to 19.15 U.T.

Colongitude 10°.7

7¼" Spec. × 292

{Seeing: 5
{Transp.: 4

Geoc   {L = +0°.56
Libr   {B = +5°.07
(19ʰ)

← Developments
since 19.15 u.T.

Shadow-profile on E. half of
interior at 21.15 u.T.   Col. 11°.8

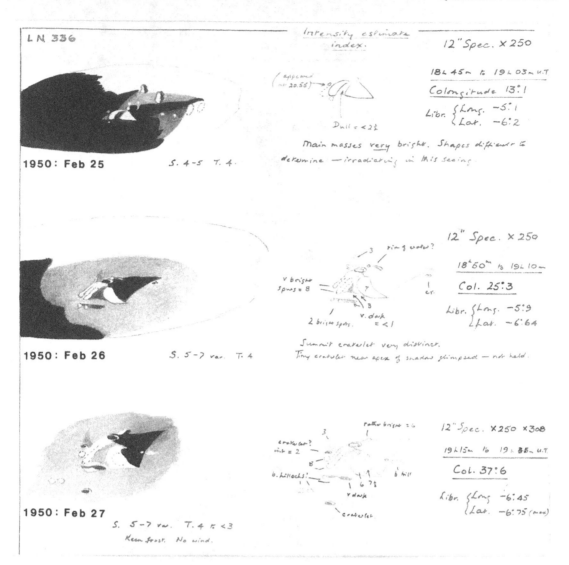

LN 336

Intensity estimate index.

12" Spec. × 250

18ʰ 45ᵐ to 19ʰ 03ᵐ U.T

Colongitude 13°·1

Libr. { Long. −5°·1
{ Lat. −6°·2

Dull = <2½

Main masses very bright. Shapes difficult to determine — irradiating in this seeing.

1950: Feb 25   S. 4–5 T. 4.

12" Spec. × 250

18ʰ 50ᵐ to 19ʰ 10ᵐ

Col. 25°·3

Libr. { Long. −5°·9
{ Lat. −6°·64

v. bright spots = 8
2 bright spots
ring crater?
v. dark = <1
cr.

Summit cratelet very distinct. Tiny cratular near apex of shadow glimpsed — not held.

1950: Feb 26   S. 5–7 var. T. 4

12" Spec. × 250 × 308

19ʰ 15ᵐ to 19ʰ 35ᵐ U.T

Col. 37°·6

Libr. { Long. −6°·45
{ Lat. −6°·75 (max)

rather bright = 6
cratular?
int = 2
b. hillocks!
b. hill
v dark
6 7½
cratelet

1950: Feb 27   S. 5–7 var. T. 4 k <3
Keen frost. No wind.

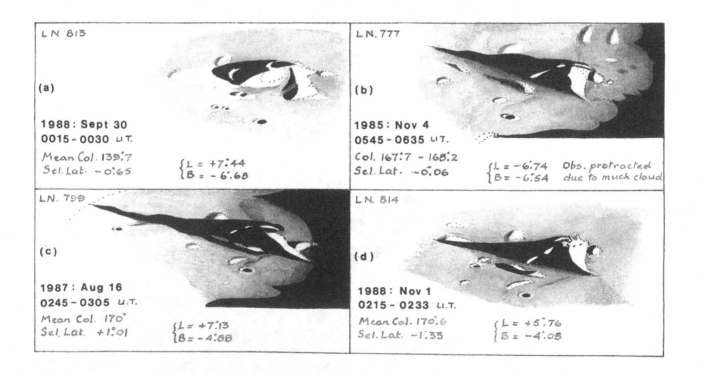

LN 813

(a)

1988: Sept 30
0015 – 0030 U.T.

Mean Col. 139°·7
Sel. Lat. −0°·65

{ L = +7°·44
{ B = −6°·68

LN. 777

(b)

1985: Nov 4
0545 – 0635 U.T.

Col. 167°·7 – 168°·2
Sel. Lat. −0°·06

{ L = −6°·74   Obs. protracted
{ B = −6°·54   due to much cloud

LN. 799

(c)

1987: Aug 16
0245 – 0305 U.T.

Mean Col. 170°
Sel. Lat. +1°·01

{ L = +7°·13
{ B = −4°·88

LN. 814

(d)

1988: Nov 1
0215 – 0233 U.T.

Mean Col. 170°·6
Sel. Lat. −1°·33

{ L = +5°·76
{ B = −4°·08

The Weigel Region

Returning to a region after the lapse of a number of years can sometimes result in new light being thrown upon a subject whose potential for further study might have been regarded as exhausted.

An example of this is the Weigel region, first observed in only a general way in early 1960 (see first drawing) and rather randomly on a number of subsequent occasions. It was not until 1984 Feb.12 that I noticed a previously unrecognised feature forming part of the lower eastern slopes of the wall which runs northwards from Weigel and which also forms the bordering rampart of a large unnamed thalassoid to the west. At the time, this gave the impression of being a lava sheet which had poured through a gap in the wall and then solidified forming an elevated plateau or terrace.

Drawings made on that date and reproduced here as (1) and (1a) show the aspect described at 2120 and 2235 UT (colongitude 40°.8 and 41°.4 respectively) as the terminator advanced westward. The appearance was reminiscent of the Grimaldi type of terracing (see Grimaldi, under Sect.11 Region 16) and excited immediate interest. A confirmatory observation was obtained two lunations later when the terminator was similarly placed. This was on 1984 April 11 2205–2245 UT colongitude 39°.8–40°.1 and drawing (2) is an extract from this observation; it shows another

'lava flow', similar in character to the south on the same 'coastline'. The narrower dark strip running concentric to the main terrace at a higher level appears on (1), (1a) and (2) but it is not clear whether this indicates merely a shallow valley or an overlapping shelf.

Observations later in the same year showed the expected reversal effect under opposite lighting and the record of 1984 Aug.21 made at 202°.8 which is shown in drawing (3) left little doubt about the true configuration in which the dipping slopes are darkening rapidly – intensity 1 – under the setting Sun, whilst the surface of the terracing at intensity 2 is starting to be engulfed in the heavy shadow from the upper wall to the west. Yet another sighting was secured on 1986 Aug.29 which again showed the reversed effect, but at colongitude 199°.6 this was less pronounced.

Reviewing the scale of the topography, it is unrealistic to suppose that these shelving, terrace-like features are indeed lava sheets. They are much more likely to be the result of down-faulting produced by the slumping of the surface along the 'coastline'. The rapid darkening of the slopes of the fault-blocks suggests an angle of inclination of the order of 15° – a not unreasonable figure taking into account the degree of thermal erosion with production of scree over aeons of time.

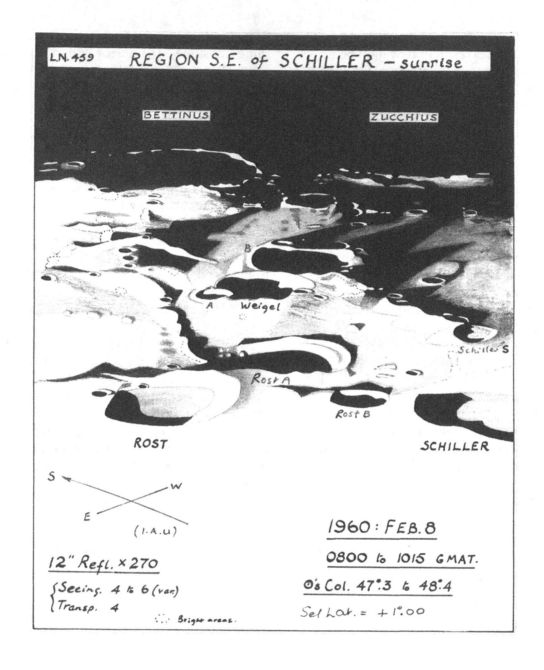

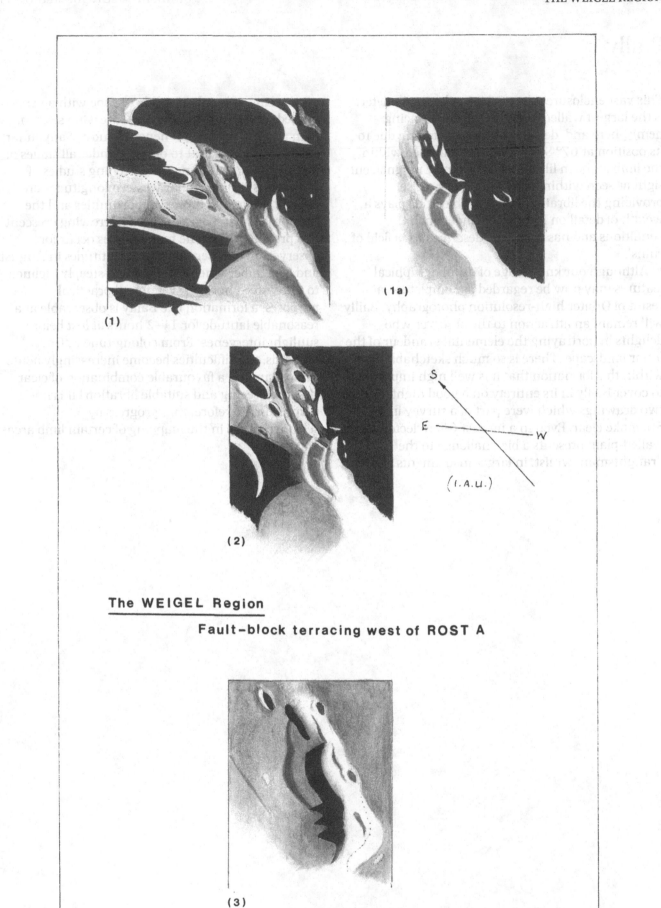

The WEIGEL Region

Fault-block terracing west of ROST A

Bailly

This vast enclosure, almost 190 miles in diameter, is the largest walled-plain on the Earth-facing hemisphere and, despite the foreshortening due to its position at 67° S 69° W (just contained within the limb at mean libration), it is a truly magnificent sight as seen within a day or two of sunrise, providing the libration is favourable. It displays a wealth of detail on its interior under such conditions and has been well described as 'a field of ruins'.

Although our knowledge of its topographical features may now be regarded as complete as a result of Orbiter high-resolution photography, Bailly will remain an attraction to the observer who delights in portraying the elemental grandeur of the lunar landscape. There is so much sketchable detail within this formation that it is well nigh impossible to cover Bailly in its entirety on a good night, as the two drawings which were part of a survey in 1949–50, make clear. Even in a modest 6½″ reflector, this walled-plain presents a big challenge to the draughtsman, whilst, in larger instruments and good seeing, it is only possible to cope with a very limited portion during any one observing session.

In making the comprehensive study of any lunar subject it is important to observe under all stages of lighting but late afternoon or evening studies of formations situated in far western longitudes are difficult because of limited opportunities and the time problem associated with a narrowing crescent in a pre-dawn sky. The best chances occur for observers in northern temperate latitudes in August and September when the ecliptic is steeply inclined to the eastern horizon; yet, for all practical purposes, a formation like Bailly is observable at a reasonable latitude for 1½–2 hours *at best* before sunlight intervenes. From colongitude 220° onwards, the difficulties become increasingly acute even assuming a favourable combination of clear sky, good seeing and suitable libration! It is not surprising, therefore, that progress by selenographers in the mapping of certain limb areas was slow.

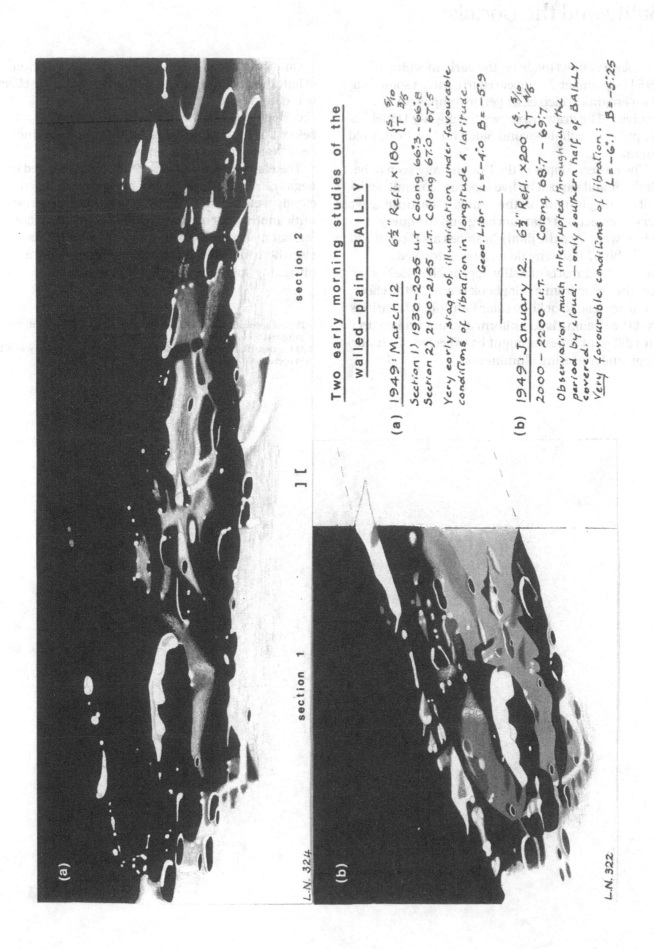

section 2

][

section 1

(a)

L.N. 324

(b)

L.N. 322.

Two early morning studies of the walled-plain BAILLY

(a) 1949: March 12   6½" Refl. × 180 {S. ⁵/₁₀ / T 3/5}

Section 1) 1930–2035 u.t Colong. 66°.3 – 66°.8
Section 2) 2100–2155 u.t. Colong. 67°.0 – 67°.5

Very early stage of illumination under favourable conditions of libration in longitude & latitude.

Geoc.Libr: L = –4°.0 B = –5°.9

(b) 1949: January 12. 6½" Refl. × 200 {S. ⁵/₁₀ / T 4/5}

2000 – 2200 u.t.   Colong. 68°.7 – 69°.7

Observation much interrupted throughout the period by cloud. — only southern half of BAILLY covered.

Very favourable conditions of libration :
L = –6°.1   B = –5°.25

Bailly and the Dörfels

The observations made on the early morning of 1951 September 27 appeared to contain something of an enigma which is, as yet, incompletely resolved. The drawing shows the mighty Dörfel range[1] on the limb beyond Bailly on the 27-day old Moon.

The eastern slopes of the Dörfels were seen to be dimly lit and capped by five separate summit peaks still shining brightly in the sunlight. Earthshine was very conspicuous that morning and the query raised was: 'Are the dimly lit slopes an effect of earthshine? If so why do not the shadowed western[2] ramparts of Bailly appear likewise?' Another note from my logbook reads: 'If the effect is due instead to oblique solar lighting, it is curious that the shading is so uniform and not broken up into different tones as might be expected of a lunar slope under grazing illumination.'

One Saros cycle later (when a precise repetition of lighting and libratory conditions occur), the effect was confirmed on 1969 October 8. On this occasion, the slopes were uniformly shaded as before but somewhat lighter in tone – earthshine being again very conspicuous.

The elapse of another 18-year period resulted in negative results for the 1987 presentation when cloudy weather intervened! There the matter rests until another opportunity offers. I have often looked for, but never seen, a remotely comparable effect elsewhere in the limb regions and it remains an intriguing subject for further investigation.

1 The old appellation for these mountains but they actually form the ramparts of Hausen and other large rings in this region.
2 At the time the notes were made the old classical orientation was still in operation.

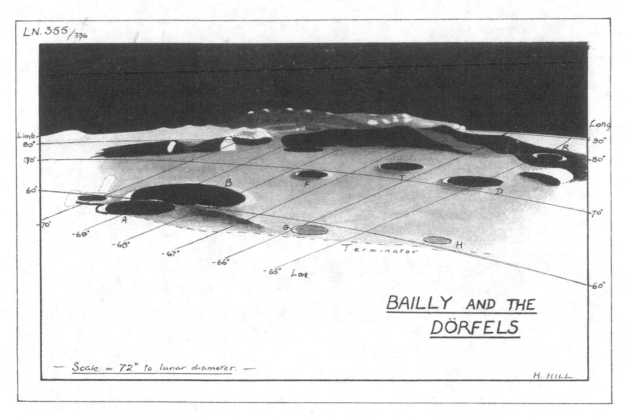

LN. 355/336

BAILLY AND THE DÖRFELS

— Scale = 72" to lunar diameter. —

H. HILL.

Instrument. 12" Spec. × 230

Seeing. —poor to fair. 3 to 5/10
transp. 3/5 Sky thinly veiled.

Could not attempt to draw the complex shadings
visible on the floor of Bailly around craters T & D
owing to the rapid approach of daybreak and unsteady air.

1951 : Sept. 27

16ʰ50ᵐ to 17ʰ50ᵐ G.M.A.T

Sun's { Colong. 239°·6 to 240°·1
Selen. Lat. −0°·64

Librations.   Long.   Lat.
Geocentric. (NA)   −4°·18   −1°·07
Corrected for parallax. −3°·57   −0°·61

Quadrant III – Section 11

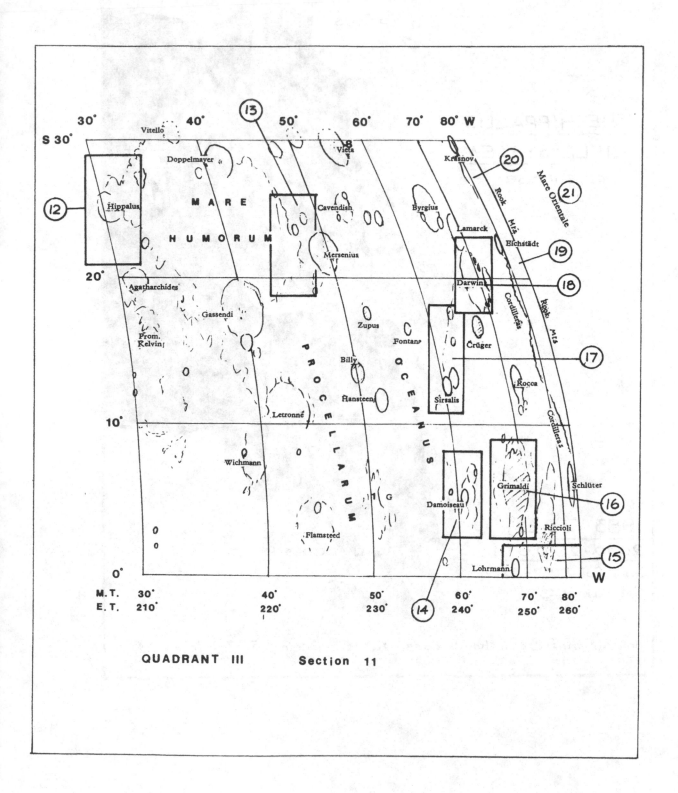

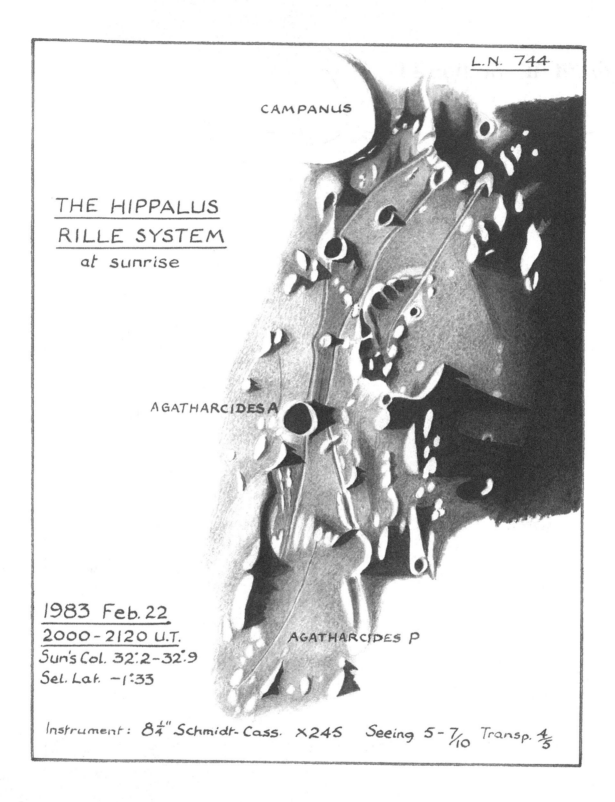

L.N. 744

CAMPANUS

THE HIPPALUS
RILLE SYSTEM
at sunrise

AGATHARCIDES A

AGATHARCIDES P

1983 Feb. 22
2000 - 2120 U.T.
Sun's Col. 32°.2 - 32°.9
Sel. Lat. −1°.33

Instrument: 8¼" Schmidt- Cass. ×245   Seeing 5 - 7/10 Transp. 4/5

The Western Shore of the Mare Humorum

The marginal zones of most of the lunar maria are regions where features resulting from subsidence and crustal movements such as wrinkle ridges, rilles and faulting are very much in evidence, and this is particularly so along the western 'shores' of the Mare Humorum. These are strikingly displayed at an early stage of sunrise.

The observation of 1983 February 23 caught the ridges and dip–slip fault lines at the critical phase when only their eastern-facing slopes are illuminated; these contrast sharply, if only briefly, with the dimly-lit or still darkened shelving regions to the west. The appearance depicted in the drawing was fleeting as in a matter of minutes it had gone, emphasising the rapid advance of sunlight.

The rilles are seen to best advantage when the morning terminator has reached the outer slopes of Mersenius (second drawing) at which time the crater enlargements along their course and the intimate relationship of these to the runs of crater-chains SE of Mersenius C are very distinct. Both rilles and faults follow a markedly circumferential or peripheral course along the 'shoreline' of Humorum.

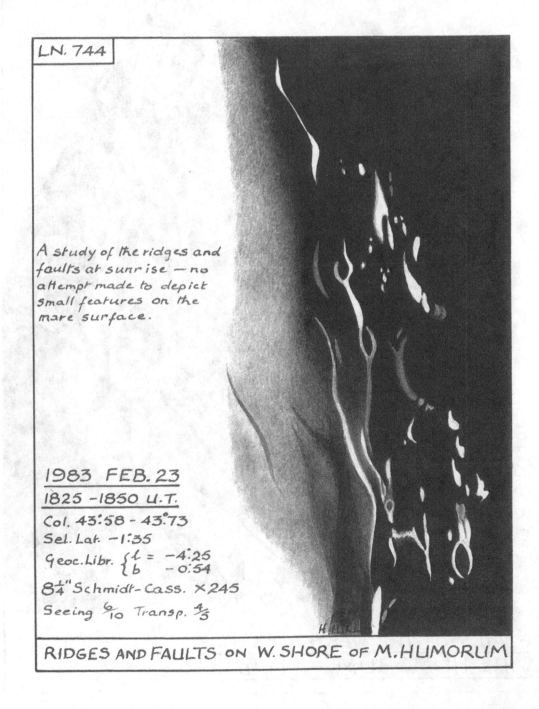

L.N. 744

A study of the ridges and faults at sunrise — no attempt made to depict small features on the mare surface.

1983 FEB. 23
1825 - 1850 U.T.
Col. 43°58 - 43°73
Sel. Lat. -1°35
Geoc. Libr. $\begin{cases} l = -4°25 \\ b\ -0°54 \end{cases}$
$8\frac{1}{4}''$ Schmidt-Cass. ×245
Seeing $\frac{6}{10}$ Transp. $\frac{4}{5}$

RIDGES AND FAULTS ON W. SHORE OF M. HUMORUM

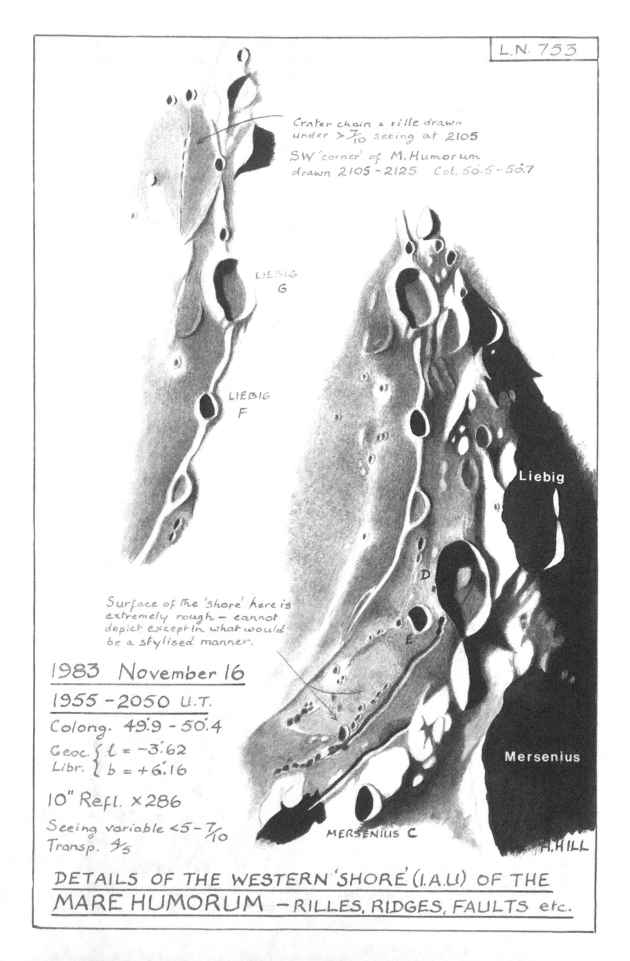

L.N. 753

Crater chain & rille drawn under $>\frac{7}{10}$ seeing at 2105

SW 'corner' of M. Humorum drawn 2105 – 2125 Col. 50.5 – 50.7

LIEBIG G

LIEBIG F

Liebig

D

E

Surface of the 'shore' here is extremely rough — cannot depict except in what would be a stylised manner.

Mersenius

MERSENIUS C

H. HILL

1983 November 16

1955 – 2050 U.T.

Colong. 49.9 – 50.4

Geoc. $\{$ $l = -3°.62$
Libr. $\{$ $b = +6°.16$

10" Refl. ×286

Seeing variable $<5 - \frac{7}{10}$
Transp. $\frac{4}{5}$

DETAILS OF THE WESTERN 'SHORE' (I.A.U) OF THE MARE HUMORUM — RILLES, RIDGES, FAULTS etc.

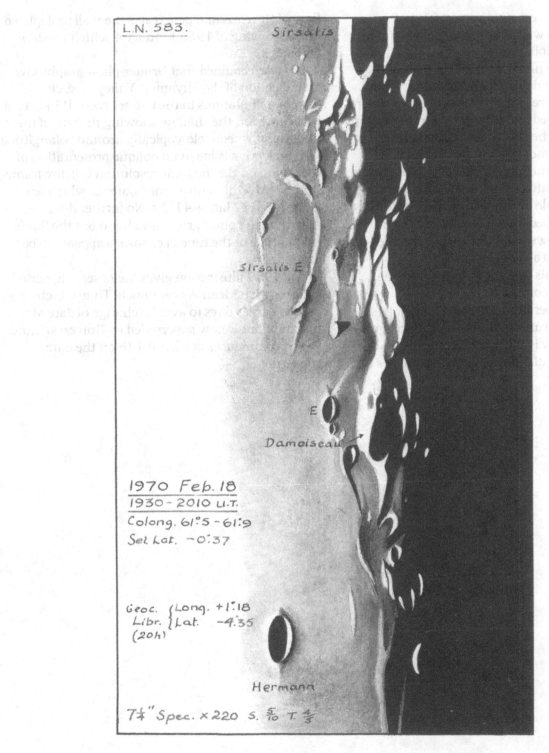

L.N. 583.

Sirsalis

Sirsalis E

E

Damoiseau

1970 Feb. 18
1930 - 2010 U.T.
Colong. 61°.5 - 61°.9
Sel. Lat. -0°.37

Geoc. {Long. +1°.18
Libr. {Lat. -4.35
(20h)

Hermann

7¼" Spec. × 220 S. 5/10 T. 4/5

**Formations on the western "shore"
of the Oceanus Procellarum**

The Miyamori Valley Region

An unadopted Japanese name given to a valley feature which runs westwards from Lohrmann to the NE wall of Riccioli. It had been recognised as a valley or rille in the nineteenth century because it appeared, variously depicted, in the maps of that period but, in more recent times, particularly during the 1950s, it received closer attention from amateur lunarians, both in the B.A.A. and elsewhere. Considerable difference of opinion remained, however, as to its true topographical configuration and, although not situated in the Moon's libratory zone, its considerable distance from disc centre did not help observers because of foreshortening effects.

The drawing shows what the author recorded in 1959 in response to a general appeal for observations and this was the first of a series he made but libratory conditions became less favourable at that period.

The narrow rille running from Hevelius towards the Miyamori Valley is the south-westerly continuation of the diagonal rille which crosses the floor of Hevelius and penetrates the wall as depicted in the drawing of 1985 January 4 which appears on page 83.

It is often claimed that Orbiter photographs give no indication of the Miyamori Valley as such, implying limitations in the Orbiter record! In actual fact, however, the shadow showing the run of the Valley as it is seen telescopically around colongitude 74°, is clearly visible in an oblique presentation of the region on the medium-resolution full disc frame IV–163–M – the centre coordinates of which are long. −53°37 lat. +41°22. No further data regarding the photograph is to hand but the Sun's colongitude at the time of exposure appears to be around 79°.

The 1959 illustration gives the observing period in Greenwich Mean Astronomical Time which was used in earlier days to avoid a change of date at midnight but is now superseded by Universal Time which commences at 0h not 12h on the date concerned.

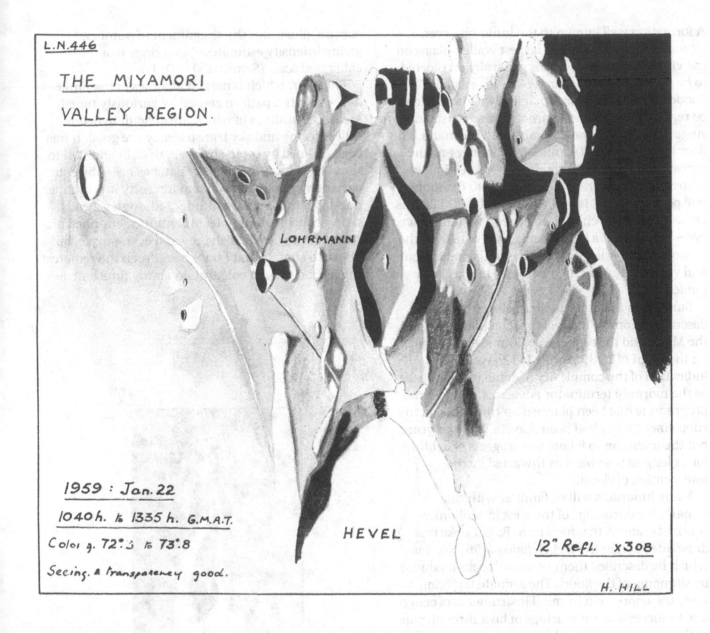

L.N.446

THE MIYAMORI
VALLEY REGION.

LOHRMANN

HEVEL

1959 : Jan. 22

1040 h. to 1335 h. G.M.A.T.

Colon g. 72°.3 to 73°.8

Seeing & transparency good.

12" Refl. x308

H. HILL

Grimaldi

A formation well known to the lunar observer, Grimaldi ranks among the largest walled-plains on the visible hemisphere and is generally considered to be one of the darkest areas – its floor being a flooded lava basin. Morphologically, Grimaldi may be regarded in size and character as a transitional stage between the walled-plains and the maria. The floor itself is about 100 miles in diameter and the surrounding ramparts measure 140×145 miles across with an average height of 4000 ft although one peak on the SE is of the order of 10 000 ft. This enclosure is large enough to be seen with the naked eye – especially if a light filter is used to take off the Moon's glare. Fuller descriptions of Grimaldi's many and varied features are to be found in the classic guides.

Sunrise on the inner western slopes has been described as one of the most spectacular sights on the Moon and the sequence of four sketches made on the night of 1987 April 11/12 gives some indication of the complexity of detail revealed here as the morning terminator advances. A revisional programme had been planned on this occasion to supplement what had been done in previous years but the intention to follow the progress of lighting for as long as possible was thwarted by the intervention of cloud.

Many lunarians will be familiar with the remarkable terracings of the west inner border south of crater A to which Dr S. R. B. Cooke first drew attention in the *B.A.A. Journal*, **56**, No. 8 in which he described them as 'fault-blocks produced by slumping of the floor'. The arcuate terracing gives the impression in small instruments of being due to successive outpourings of lava through gaps in the western wall but this argument can hardly be sustained in view of the considerable depth to the fronts.

One of the earliest observations made by the author is that dated 1948 March 23 but is very much a simplification of the true character of the terraces as in larger instruments they appear much less regular in appearance – especially in the earliest stages of sunrise. The region is to be commended to possessors of adequate aperture and observations should be conducted to follow the immediate lifting of the shadows. Owing to the position of the west wall at 71° W longitude shadow-relief is soon lost and by colongitude 75° the whole area has a 'bland' look.

The drawing captioned 'Grimaldi two days after sunrise' illustrates the assignment of numbers giving intensity estimates of markings to a calibrated scale (Schröter's) of 0–10.

The floor, which is dark at all times of the lunar day, exhibits a pattern caused by variously toned areas, the outlines of which are difficult to define unless seeing and sky transparency are good. It has been claimed by some observers that, in addition to showing colour, these dusky patches are subject to variation in both position and intensity in a manner unrelated to the solar altitude. Although green is supposed to predominate, other tints mentioned are greenish-grey, brownish grey and even purple, but I have to confess that I have never seen the remotest suggestion of such colouration at any time and, as

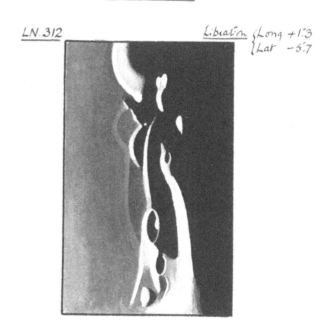

TERRACES ON E. WALL OF GRIMALDI

LN. 312

Libration { Long +1.3 { Lat −5.7

Instrument 6½" Reflector ×220

1948 March 23 8ʰ15ᵐ to 8ʰ30ᵐ GMAT. Sun's Colongitude 71°.1

{ Seeing poor = 4/10 { Transp. 4/5.

I have remarked elsewhere, the Moon, with rare exceptions, e.g. the Lichtenberg area and the 'Wood's Spot' region north of Aristarchus, always appears essentially of a neutral grey tone to my eyes.

For many years an experienced observer attributed the colour and changes he observed in his study of the dark patches within Grimaldi to the rapid seasonal growth of lowly lichenous or fungoid type of vegetation, but it would perhaps be prudent to express reservations about these claims and comment no further!

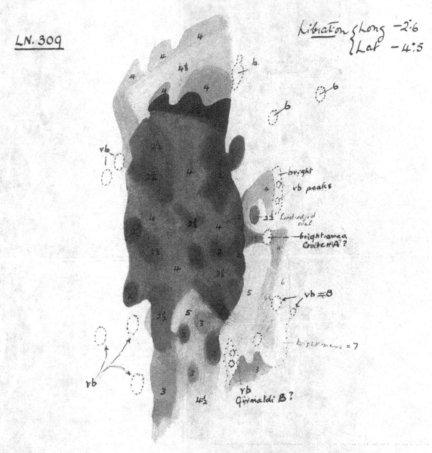

GRIMALDI
2 DAYS AFTER SUNRISE

GRIMALDI
Progress of illumination on west inner wall at sunrise

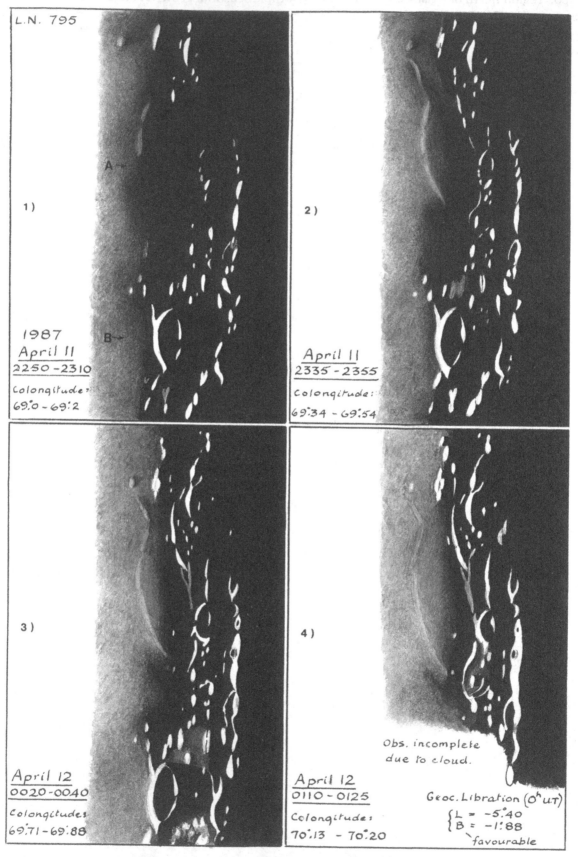

L.N. 795

1)

A →

1987
April 11
2250 –2310

Colongitude:
69°.0 – 69°.2

B →

2)

April 11
2335 – 2355

Colongitude:
69°.34 – 69°.54

3)

April 12
0020–0040

Colongitudes
69°.71–69°.88

4)

Obs. incomplete
due to cloud.

April 12
0110 – 0125

Colongitudes:
70°.13 – 70°.20

Geoc. Libration (0ʰ u.T)
{ L = −5°.40
{ B = −1°.88
favourable

— 8¼" Schmidt–Cassegrain ×245 throughout. —

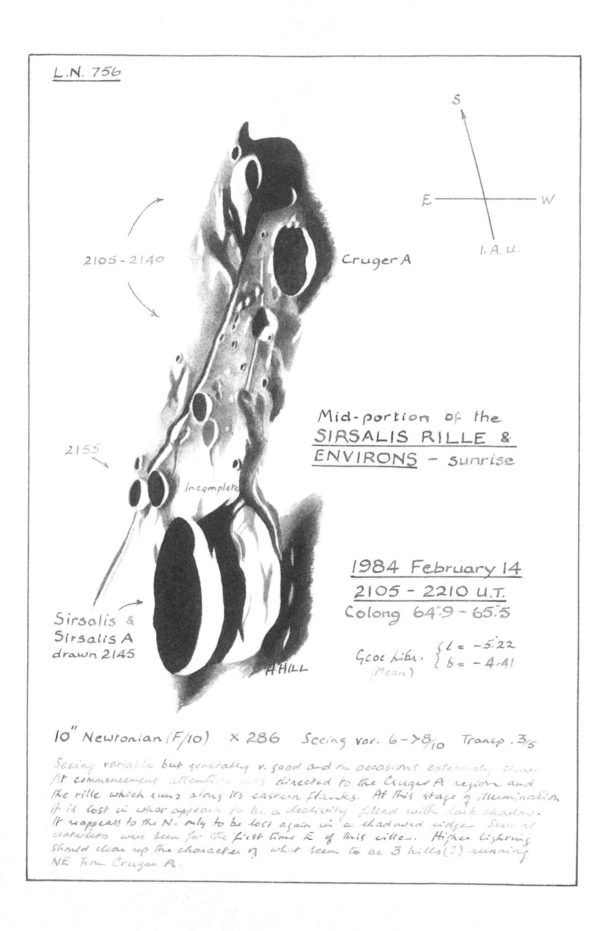

L.N. 756

S

E ——————— W

I. A. U.

2105 - 2140

Cruger A

2155

Incomplete

Mid-portion of the
SIRSALIS RILLE &
ENVIRONS - sunrise

Sirsalis &
Sirsalis A
drawn 2145

H. HILL

1984 February 14
2105 - 2210 U.T.
Colong 64°.9 - 65°.5

Geoc. Libr. $\begin{cases} l = -5°.22 \\ b = -4°.41 \end{cases}$
(Mean)

10" Newtonian (F/10) × 286 Seeing var. 6→8/10 Transp. 3/5

Seeing variable but generally v. good and on occasions extremely sharp
At commencement attention was directed to the Cruger A region and
the rille which runs along its eastern flanks. At this stage of illumination
it is lost in what appears to be a declivity filled with dark shadow.
It reappears to the N. only to be lost again in a shadowed ridge. Several
craterlets were seen for the first time E of this rille. Higher lighting
should clear up the character of what seem to be 3 hills(?) running
NE from Cruger A.

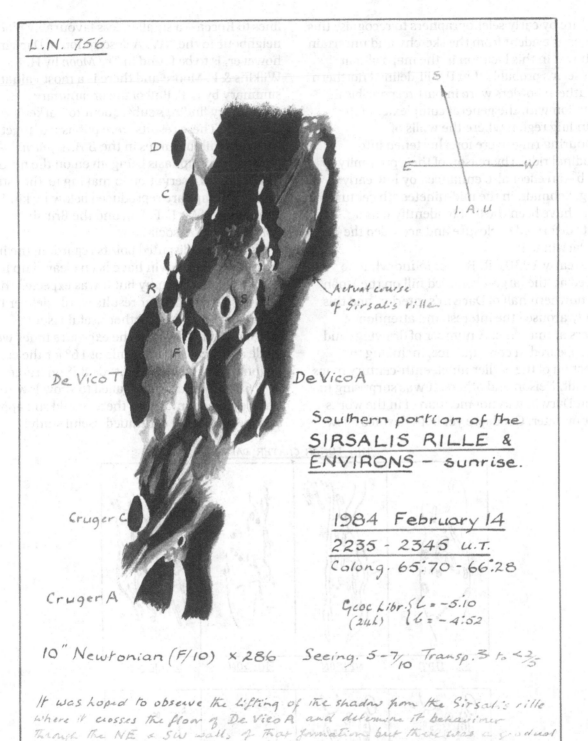

L.N. 756

S

E ———————— W

I.A.U

D

C

R

S → Actual course of Sirsalis rille

F

De Vico T

De Vico A

Southern portion of the
**SIRSALIS RILLE &
ENVIRONS** – sunrise.

Cruger C

Cruger A

1984   February 14
2235 - 2345 U.T.
Colong. 65°70 - 66°28

Geoc. Libr. $\{ L = -5.10$
(24h) $\{ b = -4.52$

10" Newtonian (F/10) × 286   Seeing. 5-7/10   Transp. 3 to <2/5

It was hoped to observe the lifting of the shadow from the Sirsalis rille where it crosses the floor of De Vico A and determine its behaviour through the NE & SW walls of that formation but there was a gradual deterioration in the transparency.

Darwin

The failure by early selenographers to recognise this formation is evident from the sketchy and uncertain lines shown in this location in the maps of that time. It seems probable that the ill-defined northern and southern borders were in part responsible in conjunction with the general complexity of the surrounding region where the walls of neighbouring rings were foreshortened into longitudinal ridges by reason of their proximity to the limb – an effect also enhanced by low early lighting. Schmidt, in the mid-nineteenth century, seems to have been the first to identify it as a separate deformed enclosure and accorded the name Darwin to it.

In the early 1930s, R. Barker found what he described as 'the largest rounded hill on the Moon' on the northern half of Darwin's interior, and this naturally aroused the interest and attention of observers at that time. A number of drawings and charts appeared in consequence, including the resurrection of the earlier nineteenth-century maps of Schmidt, Neison and others. It was surprising to find that Darwin was not mentioned in the works of Elger or, later, Goodacre yet both devoted a few lines to Rocca – a smaller, less favourably placed neighbour to the NW. A description of Darwin, however, is to be found in *The Moon* by H. P. Wilkins & P. Moore and there is a most valuable summary by L. F. Ball of the preliminary exploratory findings subsequent to Barker's discovery. These results were published, together with all available maps in the *B.A.A. Journal*, **44**/2, (1933–34), emphasis being given on the need for co-operative observation in making further studies. The eight maps are reproduced below by kind permission of Mr L. F. Ball and the British Astronomical Association.

Many of the disputed points regarding the interior topography of Darwin have been cleared up by terrestrial photography but it was expected that the high-resolution Orbiter results would deliver the coup de grâce to any further useful visual investigation. However, the exposure times were made at a local solar altitude of 16° for the central regions of Darwin (colongitude 85° on frame IV–168–H2) and Orbiter failed to show low relief features to advantage, so there would still appear to be opportunity for continued useful study.

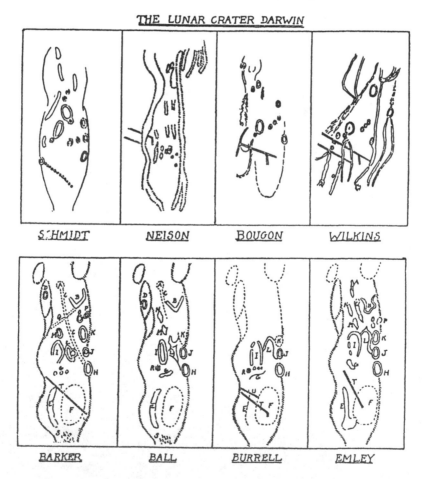

THE LUNAR CRATER DARWIN

SCHMIDT   NEISON   BOUGON   WILKINS

BARKER   BALL   BURRELL   EMLEY

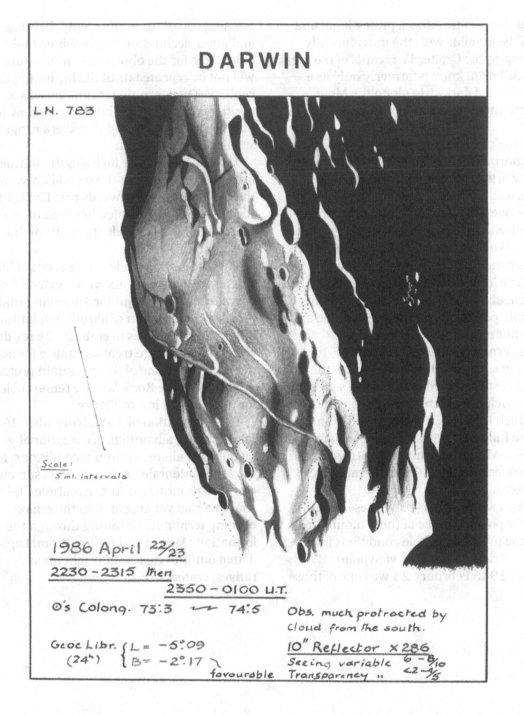

DARWIN

L.N. 783

Scale:
5 ml. intervals

1986 April 22/23

2230 - 2315 then

2350 - 0100 U.T.

⊙'s Colong. 73°3 ⟿ 74°5

Obs. much protracted by
Cloud from the south.

Geoc Libr. { L = −5°09
(24ʰ)   { B = −2°17 }
                     favourable

10" Reflector × 286
Seeing variable   6 -8/10
Transparency "   <2 -4/5

The Mare Orientale Region

All serious students of the Moon, professional and amateur, will be familiar with the magnificently detailed photographic Orbiter IV records of the great Mare Orientale basin, known formerly only as a much foreshortened dark strip along the Moon's SW limb, even under optimum librational conditions.

This is not the place to review a feature which has been so thoroughly investigated professionally, except to refer briefly to a configuration which resulted from a catastrophic event in the Moon's past. With an overall diameter of some 600 miles (though its full effects extend much further), this is the most remarkable structure of its size and type with its concentric rings and complex radial lineaments. The 'bull's eye' is a dark plain, the Mare Orientale itself, about 160 miles across, and thus relatively small, positioned on the averted side at 18° S 95°W, but marking the centre of an enormous pattern of encircling and radial mountain ranges, scarps, rilles and ridges – the youngest and best preserved of the Moon's great basins. The principal concentric rings consist of the Rook Mts, which have a double escarpment in places and are fully 380 miles in diameter; and the outer Cordillera Mts, which are some 600 miles across, thus covering some 30 selenographic degrees in extent.

The drawings which accompany these notes show what it is possible to see of these mountain rings under the most favourable conditions from our severely restricted terrestrial viewpoint.

The night of 1986 February 23 was one of those unusual occasions when seeing, lighting, libration and lunar declination all combined to the greatest advantage for the observer – circumstances which will not be repeated until the beginning of the next century. The morning terminator was ideally positioned for showing both the shadow profiles and outlines of the Rook and Cordillera ranges and the smaller ones between.

The Krasnov sector includes the southernmost limit of the Cordillera; the second drawing shows these extending northwards past Eichstadt as far as the crater Schlüter which lies west of Grimaldi at latitude 6° S and outside the limits of this presentation.

The author has made a long series of limb profile studies over many years which extend from 40° S down to the lunar equator, observing under every possible combination of libration in latitude and longitude, and hopes to embody the results of this research in a future treatise. Suffice for now to mention that some of the mountain profiles displayed by the Rook Mts are remarkable not only for their shapes but brilliance.

The observation of 1986 November 26 was made, taking advantage of exceptionally favourable, almost optimum conditions, to delineate the Mare Orientale itself against its surrounding cirque of mountains. At colongitude 204°5 no shadows had yet appeared in the region – the evening terminator running through the formations Mercator–La Hire–Prom. Laplace. The dotted outlines represent bright areas . . . mountain ranges, craters, etc.

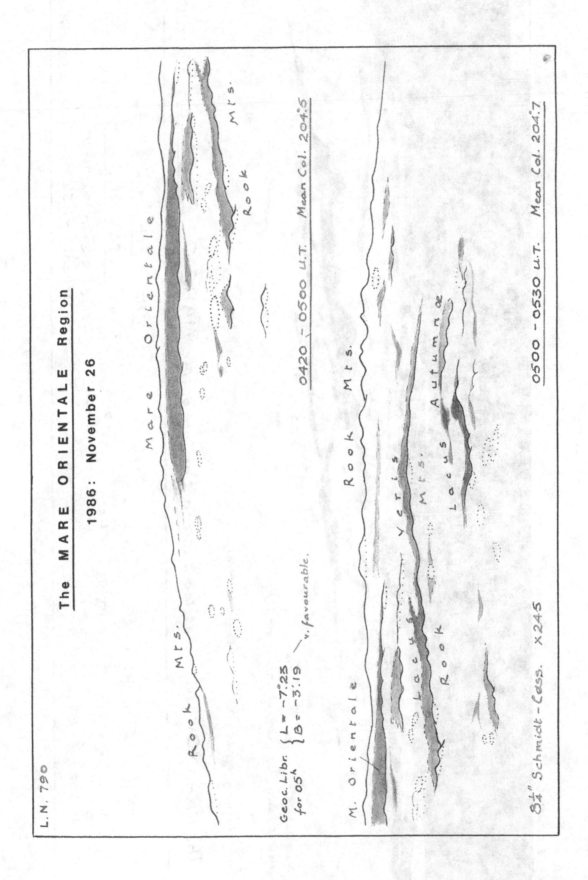

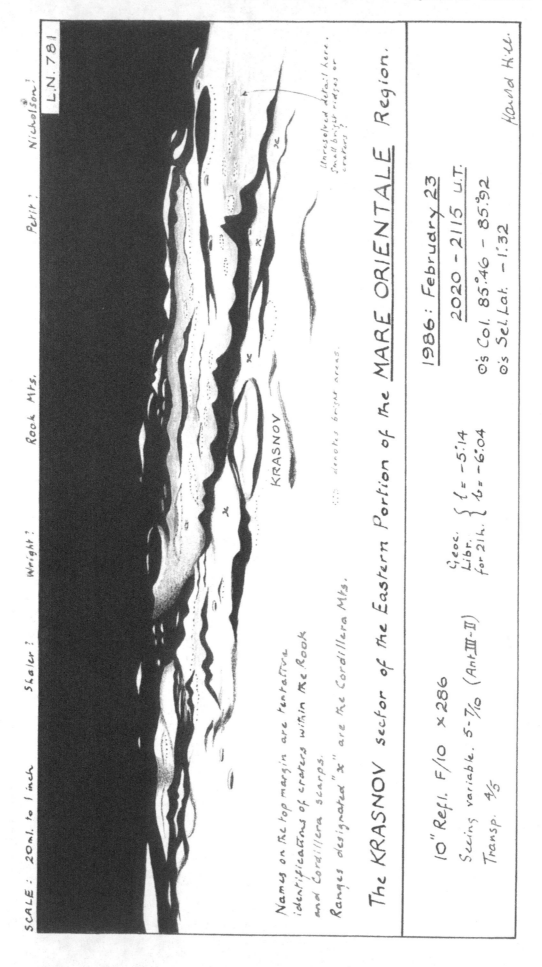

SCALE: 20 ml. to 1 inch   Shaler?   Wright?   Rook Mts.   Petit?   Nicholson?

L.N. 781

Unresolved detail here. Small bright ridges or craters?

KRASNOV

Names on the top margin are tentative identifications of craters within the Rook and Cordillera scarps.
Ranges designated "x" are the Cordillera Mts.

⟨⟩ denotes bright areas.

The KRASNOV sector of the Eastern Portion of the MARE ORIENTALE Region.

10" Refl. F/10 ×286
Seeing variable. 5-7/10 (Ant.III-II)
Transp. 4/5

Geoc.
Libr. $\{ \begin{array}{l} l = -5°.14 \\ b = -6°.04 \end{array}$
for 21h.

1986: February 23
2020 - 2115 U.T.
⊙'s Col. 85°.46 - 85°.92
⊙'s Sel. Lat. -1°.32

Harold Hill.

L.N. 781

Double escarpment of the Rook Mts.

Towards Mare centre ↓

Kopff?

SCALE: 20 ml. to 1 inch

Double escarpment of the Rook Mts.

EICHSTÄDT

The Cordillera Mts.

The EICHSTÄDT sector of the E. Portion of the M. ORIENTALE
— region at sunrise —

1986: February 23
2205 – 2250 U.T.
ō's Col. 86°.20 – 86°.71
ō's Sel. Lat. = –1°.32

10" Refl. F/10 × 286
Seeing variable 5–7/10 (Ant. III–II)
Transp. 4/5

Geoc. Libr. for 22h. { λ = –5°.13
                       G = –6°.02

Harold Hill

Quadrant III – Section 12

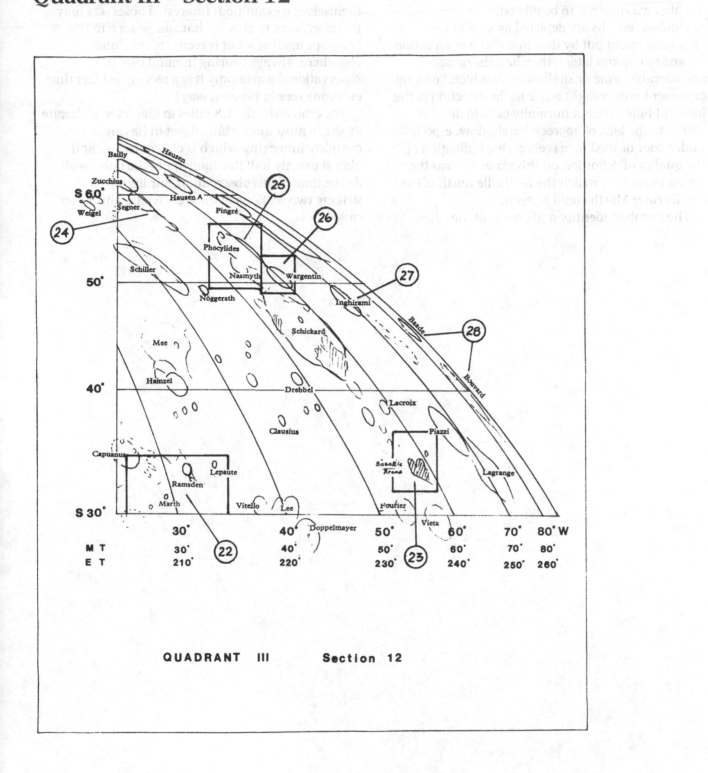

QUADRANT III   Section 12

The western portion of the Palus Epidemiarum

Although the Ramsden rille system was not on my observing list, the exceptional opportunity presented on 1989 January 17 as regards the seeing quality , was one not to be missed.

Shadow lengths are depicted as seen at the commencement but by the close of the observation – some two hours later – they had shortened considerably. Fine or shallow rilles which lie along, or present only a slight angle to the direction of the incident light, are not normally easy to discern owing to the lack of appreciable shadow, especially under poor or mediocre seeing. One indication of the quality of definition on this occasion was the distinctness with which the E–W rille north of the small crater Marth could be seen.

The number, identity and course of the rilles depicted show differences in some instances with those on published charts but it should be pointed out that the latter show disagreement among themselves if examined! Interested observers may, therefore, care to check what can be seen in the telescope against what is recorded here and elsewhere, always bearing in mind that in observational astronomy it is a recognised fact that everyone sees in his own way!

The crater Marth, 3.8 miles in diameter is, despite its size, a most interesting object in having a complete inner ring which is concentric to, and almost exactly half the diameter of the outer wall. At the time of this observation, sunlight was striking two sections of the west wall of this inner ring.

THE WESTERN PORTION OF THE PALUS EPIDEMIARUM

— morning illumination

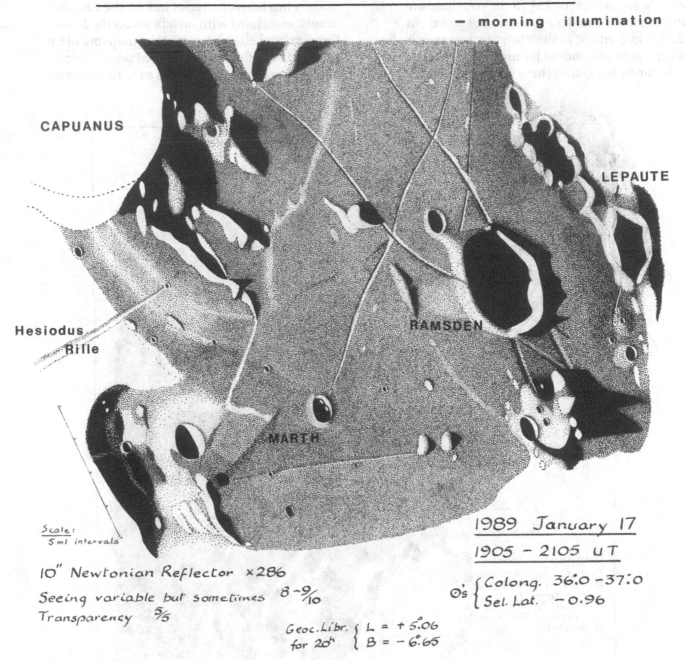

CAPUANUS

LEPAUTE

Hesiodus
Rille

RAMSDEN

MARTH

Scale:
5 ml intervals

10" Newtonian Reflector ×286

Seeing variable but sometimes $8 - \frac{9}{10}$

Transparency $\frac{5}{5}$

1989 January 17

1905 - 2105 uT

O's $\begin{cases} \text{Colong. } 36\overset{.}{.}0 - 37\overset{.}{.}0 \\ \text{Sel. Lat. } -0.96 \end{cases}$

Geoc.Libr. $\begin{cases} L = +5\overset{.}{.}06 \\ B = -6\overset{.}{.}65 \end{cases}$
for 20^h

155

The basaltic areas south of Vieta

Relatively small patches of basaltic lava are situated at a mean position of 33½° S 58° W, these are inconspicuous under a low Sun but stand out darkly in contrast to their bright surrounds under a high light and indeed for most of the lunar day. It is hardly likely that these areas will have received close attention from the telescopist. However, interest has been centred of late on the detection of 'fronts' associated with lava flows on the Mare Imbrium and elsewhere, and the mapping of the extent of such flows has assumed selenological importance in providing clues as to their origin.

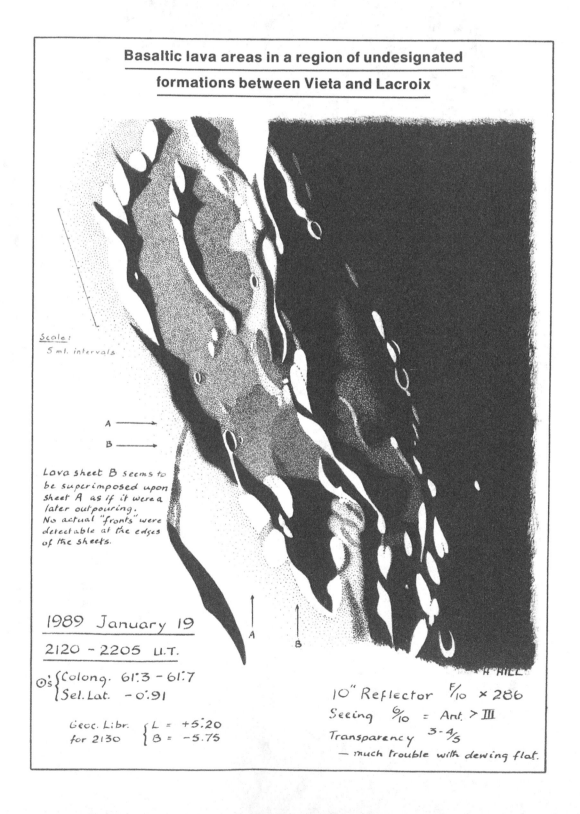

Basaltic lava areas in a region of undesignated formations between Vieta and Lacroix

Scale:
5 ml. intervals

A ⟶

B ⟶

Lava sheet B seems to be superimposed upon sheet A as if it were a later outpouring.
No actual "fronts" were detectable at the edges of the sheets.

A

B

1989 January 19

2120 - 2205 u.T.

⊙'s {Colong. 61°3 - 61°7
     {Sel. Lat. -0°91

Geoc. Libr. {L = +5°20
for 2130   {B = -5°75

10" Reflector F/10 × 286

Seeing 9/10 = Ant. > III

Transparency 3-4/5

— much trouble with dewing flat.

H HILL

156

The dark areas which lie south of the crater Vieta are clearly basaltic outflows, and the object of the author's studies in the late eighties has been to determine, if possible, the source of their origin, their effects on the neighbouring topography, and to look for small detail on their surfaces – craterpits, wrinkle ridges and other features.

It soon became evident that opportunities for low-sun observations had a severe time-limitation because of shadow from encircling elevations overlooking the site. The two drawings reproduced here were made under local solar altitudes of $3\frac{1}{2}°$ (early morning) and 14° (afternoon) from which it is obvious that low-relief objects must be sought for around colongitude 60° and not later than colongitude 230° respectively.

Whilst space photography has revealed a sprinkling of craterlets of the order of $1-1\frac{1}{2}$ km in diameter on the lava sheets, objects of this size are well beyond the capabilities of all but the largest telescopes. However, the detection of the existence of any extended curvilinear features such as lava-fronts and ripple ridges remains an open challenge for the observer who commands reasonably large apertures and any results would have a useful bearing on the purpose of the enquiry outlined here.

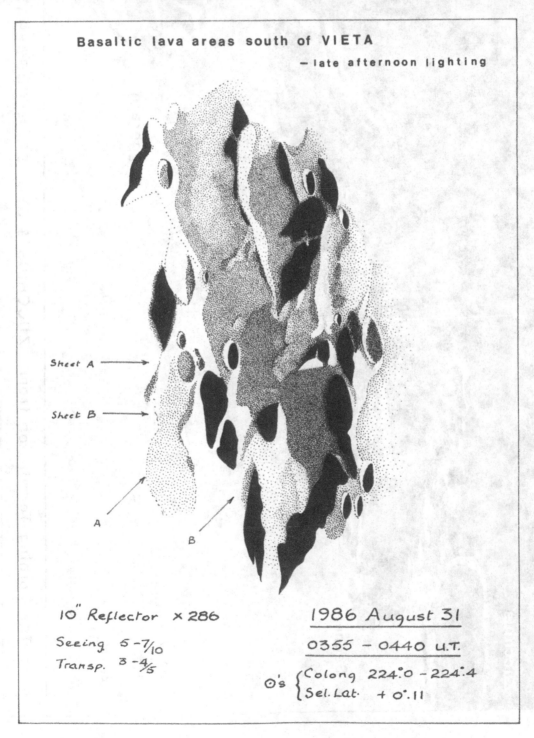

Basaltic lava areas south of VIETA
– late afternoon lighting

Sheet A →
Sheet B →
A
B

10" Reflector × 286

Seeing 5 -7/10
Transp. 3 -4/5

1986 August 31

0355 - 0440 U.T.

☉'s { Colong 224°.0 - 224°.4
{ Sel. Lat. + 0°.11

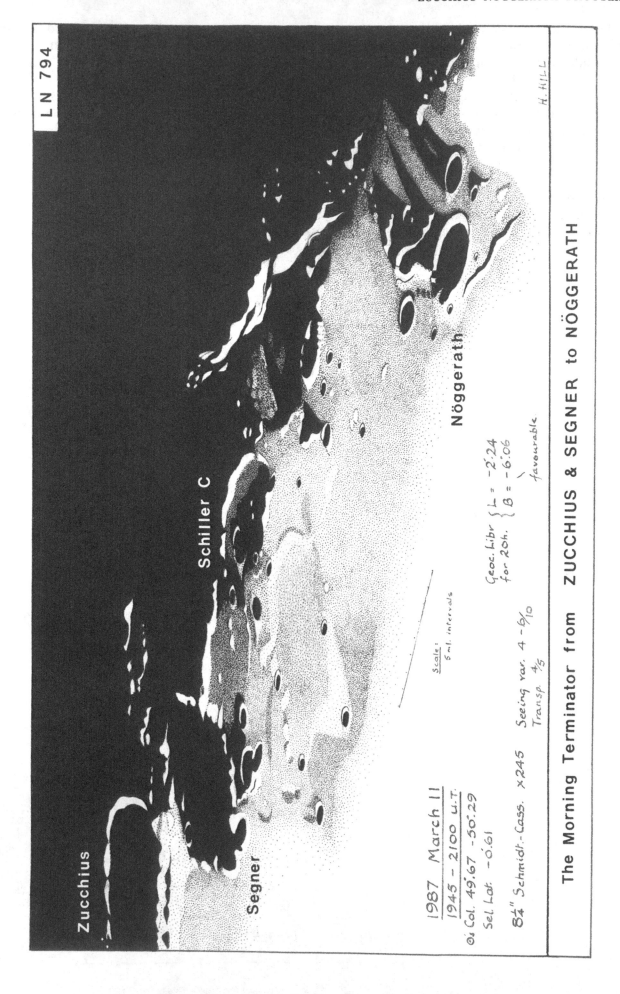

LN 794

Zucchius

Schiller C

Nöggerath

Segner

H. HILL

1987 March 11
1945 – 2100 u.T.
⊙ Col. 49°.67 –50°.29
Sel. Lat. –0°.61

8¼" Schmidt-Cass. ×245   Seeing var. 4 – 6/₁₀
                                              Transp. 4/5

Scale:
5 ml. intervals

Geoc. Libr. { L = –2°.24
for 20h. { B = –6°.06
}
favourable

The Morning Terminator from   ZUCCHIUS & SEGNER to NÖGGERATH

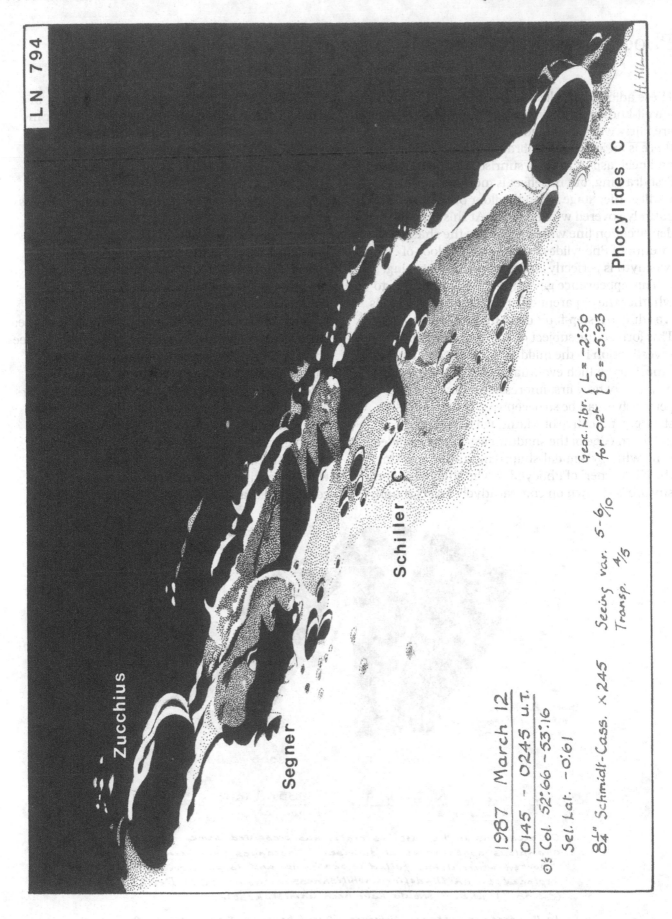

LN 794

Zucchius

Segner

Schiller C

Phocylides C

1987 March 12
0145 – 0245 u.t.
0ˢ Col. 52°.66 – 53°.16
Sel. Lat. –0°.61

8¼" Schmidt-Cass. ×245

Geoc. Libr. { L = –2°.50
            { B = –5°.93
for 02ʰ

Seeing var. 5–6/10
Transp. 4/5

H. Hill

Phocylides and Nasmyth

These adjoining walled-plains form with Wargentin a well-known group in the SW quadrant and they are fairly well positioned for observation unless there is an adverse NE libration. The familiar 'sole and heel' aspect at early sunrise is evident in the first drawing, but the effect is more pronounced at a still earlier stage when the floor of Phocylides is entirely covered with shadow. At this time the demarcation line which separates the shadowed interior of Phocylides from the sunlit floor of Nasmyth is perfectly straight and most striking.

This appearance gave rise to the question as to whether the apparent difference in the floor levels was due to a step-fault dividing the two formations. This formed the subject of an observational investigation in the middle 1950s by amateur lunarians[1] which eventually proved to be a classic instance of how first impressions, however persuasive, can be so deceptive. The straight line, strongly indicative of a fault, was in fact the northern edge of the shadow cast by the rounded, somewhat pyramidal-shaped mountain marking the NE 'corner' of Phocylides. The shadow at early sunrise is thrown on comparatively level ground,

but the subsequent shortening and movement in azimuth of this shadow as the Sun rises higher in the lunar sky, causes it to fall on the more broken terrain comprising the dividing-wall proper, thus destroying the illusion of a step-fault. A facile enough explanation when viewed in retrospect, perhaps, but one which required a careful analysis of many observations to elucidate fully.

Mention must be made of the occasional anomalies aspect of the principal floor crater in Nasmyth which lies just SW of centre. This could not be found at colongitude 59° on 1982 Dec. 27 – its place being occupied by an ill-defined whitish patch at the nearest point of the rhomboidal-shaped shadow covering the SW portion of the interior (see drawing). A little research soon uncovered an earlier observation (undated) by M. Cave, who failed to record Nasmyth A whilst including craters of lesser size on the interior. A similar anomalous observation was made by P. Wade on 1981 Dec. 8 in which a bright patch only marked the crater's position with dusky shading from it to the west wall. In seeking for an explanation it was considered that the abnormality might be due to

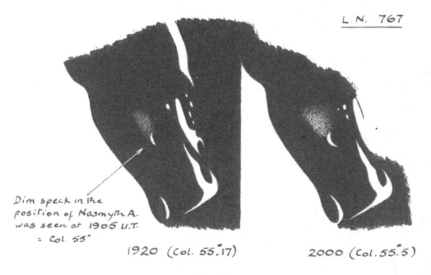

L.N. 767

Dim speck in the position of Nasmyth A. was seen at 1905 U.T. = Col. 55°

1920 (Col. 55.°17)   2000 (Col. 55.°5)

Appearance of Nasmyth A at sunrise

1985: January 3
O's Sel. Latitude = −1.°13

On occasions in the past this crater has presented some anomalous appearances at sunrise. Instances have been reported where it has failed to be visible and its position replaced by an ill-defined whitishness — as on 1982: Dec. 27 (Col. 58.°7) when it should have been unmistakeable.

In the present instance, however, it will be seen that Nasmyth A displayed a perfectly normal aspect and its crateriform aspect was evident from 1920 U.T. onwards.

Instrument: 10" Reflector ×190 { Seeing < 5/10 - deteriorating Transp. 3/5   (Ant III)

some trick of lighting and that perhaps the crater lies on a west-facing slope and thus unfavourably placed to receive the first solar rays. However, the fact that it was well seen on 1985 Jan.3 at colongitude 56° as a distinct crateriform object negates this supposition. In the absence of any other explanation, we have to regard this as a case of occasional obscuration, though from what agency remains to be revealed.

The second drawing made on 1983 Oct.2 under afternoon lighting shows Nasmyth A as a prominent 'normal' object with a smaller companion craterlet on its NE flank. This observation is also included to show the ease with which floor craterlets can be made out under 'contre-jour' lighting because of the greater areas of shade presented under such conditions.

1 The observational results and various viewpoints put forward were published in the B.A.A. periodical 'The Moon' during the editorship of Frank Thornton – the relevant numbers being found in Vols.3–5 inclusive.

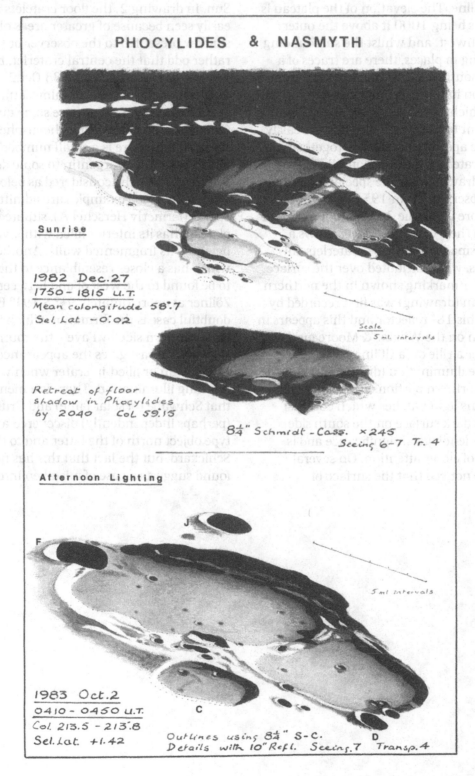

PHOCYLIDES & NASMYTH

Sunrise

1982 Dec.27
1750 – 1815 U.T.
Mean colongitude 58°.7
Sel. Lat -0°.02

Retreat of floor shadow in Phocylides by 2040   Col 59°.15

Scale
5 ml intervals

8¼" Schmidt-Cass. ×245
Seeing 6-7 Tr. 4

Afternoon Lighting

5 ml intervals

1983 Oct.2
0410 – 0450 U.T.
Col. 213.5 – 213°.8
Sel. Lat. +1°.42

Outlines using 8¼" S-C.
Details with 10" Refl. Seeing.7 Transp.4

Wargentin

Positioned at 51° S 60° W in the SW quadrant, Wargentin is well known on account of its almost unique character in giving the appearance of having been filled to the brim with lava. It is some 54 miles in diameter (or rather less than Copernicus) but, as a result of foreshortening, it resembles a thin or shallow oval plateau, somewhat polygonal in outline. The elevation of the plateau is generally cited as being 1000 ft above the outer plain to the northwest, and whilst the surrounding rampart is missing in places, there are traces of a low wall on the south-west. Wargentin may seem devoid of detail on its surface, but a low sun reveals several ridges which have been aptly likened in their arrangement to a crow's foot. They are easily seen in moderate apertures at sunrise together with a sprinkling of craterlets if seeing conditions are favourable (see drawing) but the special map prepared from observations in 1952 April by Wilkins and Moore using the 33″ Meudon refractor and published in their work *The Moon*, shows a wealth of much smaller features – craterlets, low hills and mounds, well distributed over the surface. The delicate linear marking shown in the northern part of the plateau (drawing) was first recorded by Thornton using his 18″ reflector and this appears in the same position on the Wilkins & Moore map. It seems to be either a rille or fault line.

Under opposite illumination (drawing 2) the segment of the northern portion enclosed by this rille/fault contains light patches which contrast sharply with the dark surface on the south side. This may have selenological significance and is perhaps worthy of closer attention. On several occasions I have noticed that the surface of

Wargentin looks much brighter under afternoon lighting than soon after sunrise – the effect is shown in the illustrations – an unexpected difference if the small-scale texture of the plateau is rugged. Under afternoon lighting, extensive light areas appear on the floor and these cannot be readily associated with the topography exhibited under a low opposite Sun. In drawing 2, the floor cratelets are more easily seen because of greater areas of contained shadow presented to the observer but it seems rather odd that the central craterlet, though looked for, was not recorded on 1983 Oct.2.

The epithets 'unique' or 'almost unique' as applied to Wargentin require some qualification. Careful search of the visible hemisphere has revealed that there is a small number of objects which resemble Wargentin to some degree – sufficiently so to be considered as belonging to the same class. Good examples are admittedly few but Spörer (formerly Herschel A), situated on a circular plateau, has its interior filled with lava almost to the height of its fragmented walls. Another example, which has a closer resemblance to the prototype, is to be found to the west of and between Kant and Zöllner at approximately $9\frac{1}{2}$° S $19\frac{1}{2}$° E. A more doubtful case is the formation which is broken into on its western side by Faye – the rounded scarp slope on the east gives the appearance of the edge of a plateau or filled-in crater when viewed under morning illumination. There is a mention by Webb that Schwabe (of solar fame) and Gruithuisen (perhaps independently) discovered a Wargentin-type object north of the latter and to the west of Schickard, but the fact that this has never been found suggests a case of mistaken interpretation.

WARGENTIN

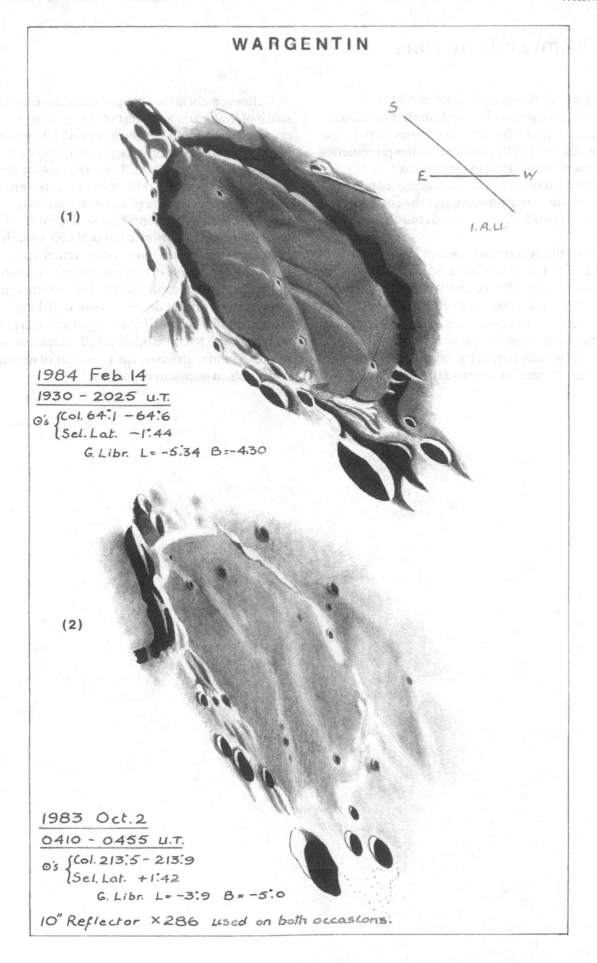

(1)

1984 Feb. 14
1930 - 2025 u.T.
⊙'s {Col. 64°.1 - 64°.6
     {Sel. Lat. -1°.44
        G. Libr. L = -5°.34 B = -4°.30

(2)

1983 Oct. 2
0410 - 0455 u.T.
⊙'s {Col. 213°.5 - 213°.9
     {Sel. Lat. +1°.42
        G. Libr. L = -3°.9 B = -5°.0

10" Reflector X286 used on both occasions.

S

E ———————— W

I.A.U.

Inghirami and environs

Situated NW of Wargentin at 48° S 69° W Inghirami is somewhat larger than that formation at 57 miles across. The difference between the relative obscurity of the former and the prominence of Inghirami is a striking example of our dependence on the interplay of highlight and shadow on the lunar topography; the contrast between the two formations could hardly be greater.

The beautifully terraced walls of Inghirami rise to some 12 500 ft above a depressed floor and a conspicuous rille-valley runs NW from the central elevations to make a broad gash in the NW rampart, continuing outside as a coarse form of 'mur énigmatique' at this stage of lighting, to eventually become part of the large broad valley which extends much further to the NW (see Region 28).

A closely associated group of sizeable craters lies south of Inghirami and west of Wargentin. A prominent rille with rugged banks which appears to consist in part of crater enlargements can be traced south as far as Phocylides. This and another shorter rille seem to be closely related to the crater-group from which they converge towards Inghirami. Having regard to the considerable degree of foreshortening, it is more than probable that these 'rilles' are quite broad, comprising crater-rows.

During the course of this particular observation, the brilliantly lit elevations which were emerging from the lunar night developed some striking configurations before they enlarged and merged with the complexity of other detail. The scene was one of extreme intricacy, the grandeur of which was impossible to adequately portray.

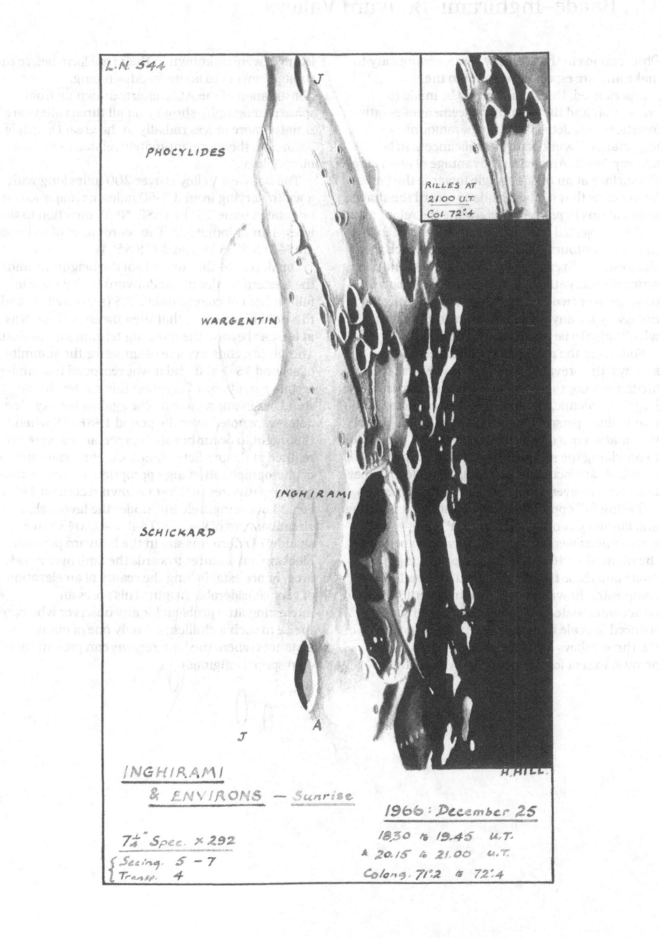

L.N. 544

J

PHOCYLIDES

RILLES AT
2100 U.T
Col. 72°.4

WARGENTIN

INGHIRAMI

SCHICKARD

J   A

INGHIRAMI
& ENVIRONS — Sunrise

7¼" Spec. × 292
{ Seeing. 5 – 7
{ Transp. 4

H. HILL.

1966: December 25
18.30 to 19.45 U.T.
& 20.15 to 21.00 U.T.
Colong. 71°.2 to 72°.4

The Baade–Inghirami–Bouvard Valleys

Observations in the libratory zones are not easy to make and are especially difficult to the inexperienced. Due regard has to be made to proportion and the accurate placement of greatly foreshortened detail; this is of paramount importance if work of any significance is to be accomplished. An obvious advantage of observing the surface at an oblique angle towards the limb, however, is that it enables one to record the shapes of elevations in *profile*, to determine the relationship to their projected shadows, and it can reveal clues as to the contours of the surface upon which the shadows fall. Even so, difficulties abound and a long series of observations of a given area – extending perhaps over two or more Saros cycles – may be necessary for any interpretational ambiguities which arise to be satisfactorily resolved.

Following the results of the high-resolution surveys, the prevailing opinion has been that any further selenographic mapping aspirations are now largely academic. Against this view it has to be said that oblique perspective work can still yield useful information of a complementary nature, in addition to providing the serious investigator with much personal satisfaction in what it is possible to achieve from even a severely restricted terrestrial position.

Taking full opportunity of advantageous lighting and libratory conditions at the SW limb, observations were carried out over a 3 h period on the night of 1986 March 24. Two drawings were made and these have been joined to form the composite, shown opposite, but in order for them to be accommodated here they have been much reduced in scale from the originals. The positions of the three valleys are indicated by the marginal arrows. Except for the Inghirami Valley, these features were unknown in their true form before the Orbiters, owing to acute foreshortening. Consultation of the ACIC charts drawn up from space photography shows that all three valleys are situated more or less radially to the great Orientale basin and, therefore, intimately related to its morphology.

The Bouvard Valley is over 200 miles long with a width varying from 35–50 miles, its major axis is orientated some 30° in a SSE–NNW direction to the lines of lunar longitude. The coordinates of its limits are: 43° S 80°.5 W and 33° S 85° W.

On March 24 the summits of the heights forming the western border of the Bouvard Valley were in full sunlight at colongitude 79°.5 (see drawing) and this notwithstanding that their mean position was at least 4° beyond the morning terminator position. This gives a conservative estimate for the summits of around 15 000 ft and it was reasoned that under certain conditions of libration this border should stand out in impressive profile against the sky. Ten dates were noted over the period 1986–88 when this ought to occur but such expectations were not realised at the predicted times. Careful examination of photographs showing appropriate librations also gave negative results! Yet my own record on 1986 Feb.23 at colongitude 86° under the favourable circumstances of $L = -5°.14$ $B = -6°.04$ showed a shadow of enormous size in the Bouvard position blocking out features towards the limb over a wide area, hence establishing the reality of an elevation of very considerable height. This poses an interesting little problem for any observer who cares to rise to such a challenge – only one of many instances where the limb regions can present their own special enigmas!

The BAADE INGHIRAMI & BOUVARD VALLEYS at sunrise

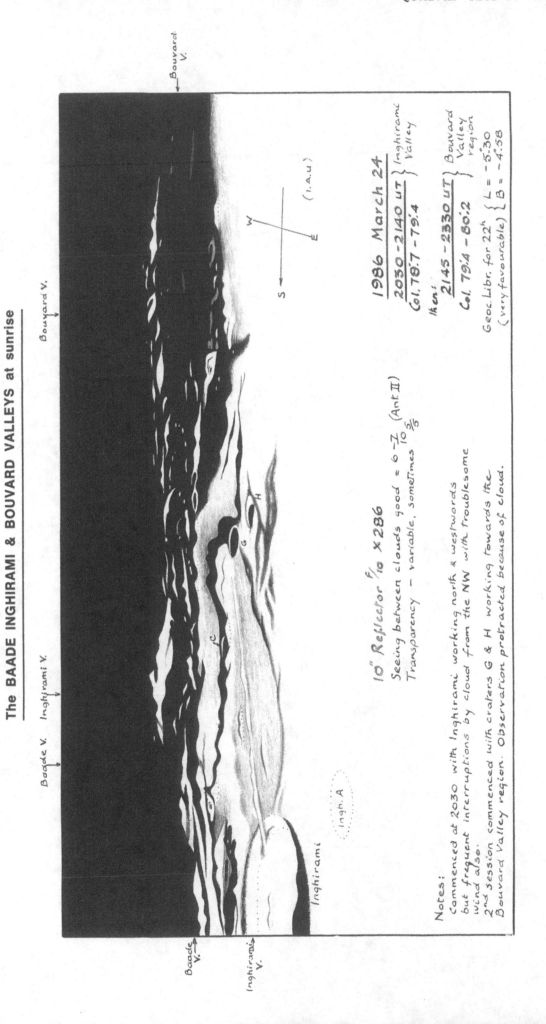

Baade V.   Inghirami V.

Bouvard V.

Bouvard V.

Inghirami

Inqh. A

Baade V.

Inghirami V.

S
W
E
(I.A.U.)

10" Reflector f/10 ×286
Seeing between clouds good = 6–7 (Ant. II) sometimes 10 5/5
Transparency – variable, sometimes 10 5/5

1986 March 24
2050–2140 UT } Inghirami
Col. 78·7 – 79·4   Valley

then:
2145–2330 UT } Bouvard
Col. 79·4 – 80·2   Valley region

Geoc. Libr. for 22ʰ { L = –5·30
(very favourable) { B = –4·58

Notes:
Commenced at 2030 with Inghirami working north & westwards but frequent interruptions by cloud from the NW with troublesome wind also.
2ⁿᵈ session commenced with craters G & H working towards the Bouvard Valley region. Observation protracted because of cloud.

Quadrant IV – Section 13

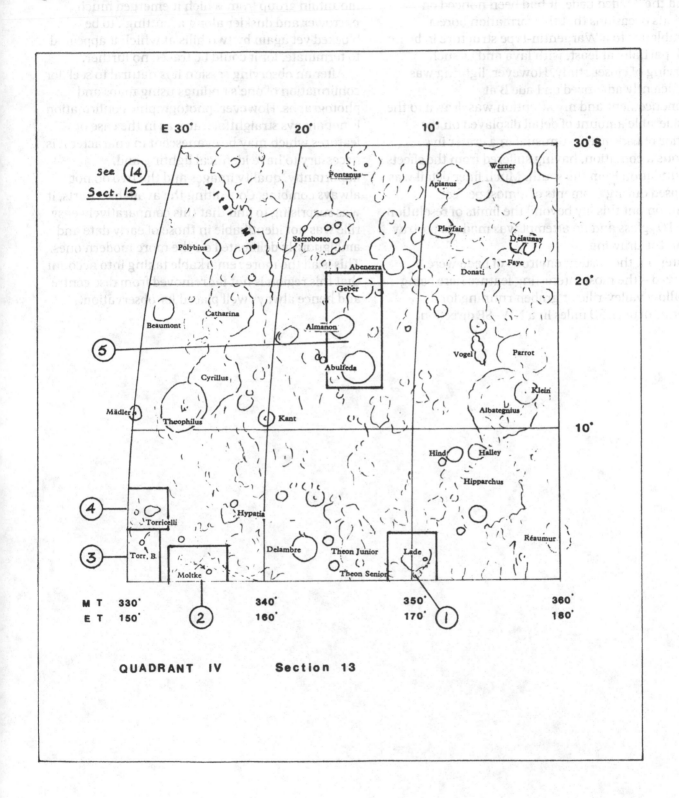

See ⑭
Sect. 15

ALTAI MS.

E 30°   20°   10°

30° S

Werner

Pontanus
Apianus

Playfair
Delaunay
Sacrobosco
Polybius
Faye
Abenezra
Donati
Geber
20°

Catharina
Almanon
Beaumont
Abulfeda
Vogel
Parrot

⑤
Klein
Cyrillus
Albategnius

Mädler
Kant
Theophilus
10°

Hipparchus
Hind
Halley

Réaumur
④
Hypatia
Torricelli

Torr. B.
Delambre
Theon Junior
Lade
③
Theon Senior
Moltke

M T   330°   340°   350°   360°
E T   150°   160°   170°   180°

②   ①

QUADRANT IV   Section 13

Lade and its environs

One of the items on my observing programme for 1989 Feb.12 was the examination of the enclosure designated Lade B which is situated between Godin A and the flooded Lade. It had been noticed on previous occasions that this formation bore a resemblance to a Wargentin-type structure in being filled, partially at least, with lava and as such deserving of closer study. However, lighting was insufficiently advanced on Lade B at commencement and my attention was drawn to the considerable amount of detail displayed on the interior of Lade itself – unusual in a crater in so ruinous a condition, having suffered from the effects of inundation from the south. Much finer detail was glimpsed during moments of almost perfect definition but this lay beyond the limits of resolution for a 10″ glass and no attempt was made to portray this in the drawing.

Later on, the eastern environs of Lade were observed – the most interesting feature here being a shallow valley-rille or graben running for a distance of over 50 miles in a NW–SE direction. Beginning in the mountain ridges east of Godin A as a broad groove with raised banks, it was interrupted in its course southwards by an irregular mountain group from which it emerged much narrower and duskier along a 'cutting', to be blocked yet again by two hills at which it appeared to terminate, for it could be traced no further.

After an observing session it is natural to seek for confirmation of one's findings using maps and photographs. However, photographic confirmation is not always straightforward, as in the case of features which may be evanescent in character it is necessary to have identical lighting and, importantly, quality images and the two do not always combine. Consulting the available charts, it was surprising to find that this comparatively easy rille was not identifiable in those of early date and ambiguously delineated on the more modern ones. This is all the more remarkable taking into account that this region is not far removed from disc centre and hence always well placed for observation.

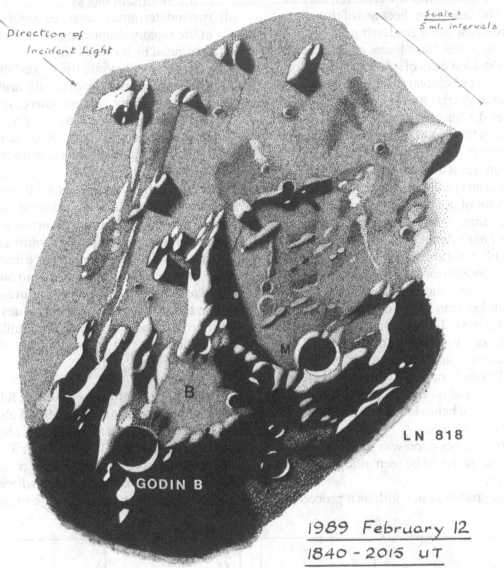

LADE AND ENVIRONS

at the Morning Terminator

Direction of Incident Light

Scale: 5 ml. intervals

M

B

LN 818

GODIN B

1989 February 12

1840 - 2015 UT

O_5 { Colong. 352° - 352°7
{ Sel. Lat -0°30

$8\frac{1}{4}$" Schmidt-Cass. ×245
& 10" Newtonian Reflector ×286

General outlines at first with driven $8\frac{1}{4}$", Then details filled in
using The 10" as seeing improved which at best reached 8-9/10
Transparency excellent throughout = 5/5

The eastern extensions of the Hypatia Rille

This drawing and accompanying notes took the form of a communication to the editor of the B.A.A. magazine *The New Moon* and they appeared in the 1987 December issue. They are repeated here in substantially the same form because of their possible interest to a wider readership.

Typical of the contentious issues raised from time to time in the 'golden days of selenography' is one which concerns the behaviour of the Hypatia rille east of the crater Moltke, and in particular from the site of craterlet B. Charts A and B (below) are reproduced from the B.A.A.'s earlier periodical *The Moon*, 9/2 January 1961 in which a forked section of the rille in question and named R1 and R2 is shown. These charts were drawn up by Dr Brian Warner (Director of the B.A.A. Lunar Section 1962–64); the first, A, was traced by Dr Warner from Kuiper's *Photographic Atlas* and the second, B, from his own observations using the 18″ refractor at the Mill Hill Observatory, London. A shows R1 and R2 ending near lunarite ridges. However, Warner concluded from his own observations that both R1 and R2 passed through gaps and proceeded beyond the ridges for some distance.

All this is relevant to an observation made on 1987 Sept. 13 (shown opposite) under excellent seeing conditions and with the lighting just perfect for investigating the behaviour of the eastern section of the Hypatia rille.

Definition on this occasion was so sharp that:

(a) strips of shadow could be seen along sections of the R1 rille;

(b) R2 tapered gradually in width as it proceeded eastwards from Moltke B and ended firmly at the foot of the southernmost mountain block;

(c) craterpits were detected on the summits of two of the mountain blocks;

(d) two indeterminate features could be discerned in the rapidly shadowing section of the area contained by R1 and R2.

Heavy shadow beyond the mountain blocks prevented any possibility of the R1 and R2 branches being traced further. One obvious difference from the charts is that I saw R1 as running directly into Moltke B whereas the diagrams show it as bypassing the craterlet on its northern side before joining the main rille.

It is curious how observers see things differently, and very often the ultimate court of appeal – a good-quality photograph – may not prove as decisive as one might wish on the precise points at issue. Lunarians with adequate aperture may care to make further observations to add to our knowledge of this area and also confirm or refute a long-held notion that the Hypatia rille continues across the southern fringes of the Mare Tranquillitatis towards Censorinus as depicted on a number of maps.

Added note The bright raised bank of R1 was beautifully seen at sunset on 1983 Feb. 3 at colongitude 153°5 as a 'mur énigmatique' – a term first used by the French astronomer Trouvelot to describe the last remnants of rilles running beyond the terminator into the darkness where they can be seen in some instances as delicate threads of light.

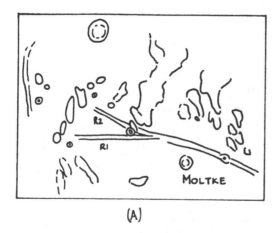

(A)

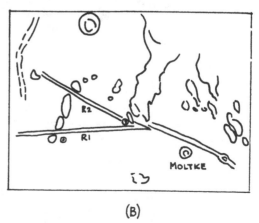

(B)

RILLE SOUTH OF MOLTKE taken from *The Moon*, 9 No.2, January 1961, p.28

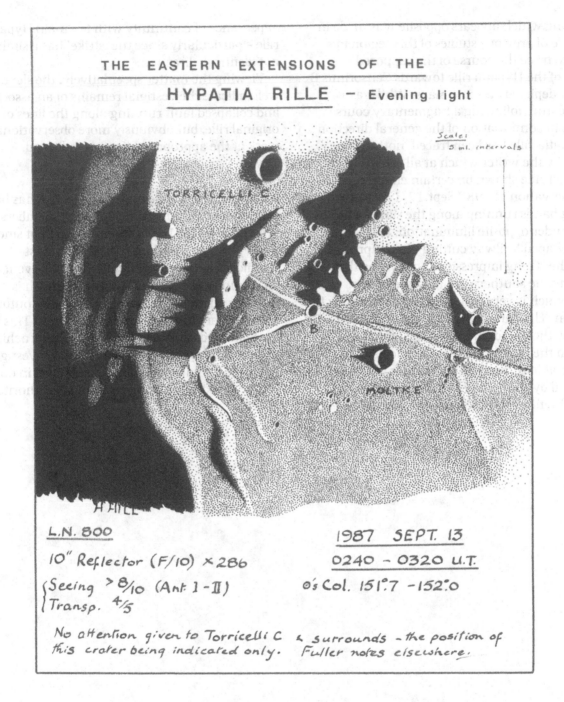

THE EASTERN EXTENSIONS OF THE

HYPATIA RILLE – Evening light

TORRICELLI C

Scale:
5 ml. intervals

B

MOLTKE

A HILL

L.N. 800

10" Reflector (F/10) ×286

(Seeing >8/10 (Ant. 1-II)
Transp. 4/5

1987   SEPT. 13

0240 – 0320 U.T.

0's Col. 151°.7 – 152°.0

No attention given to Torricelli C
this crater being indicated only.

& surrounds – the position of
Fuller notes elsewhere.

Torricelli B and environs

The drawing which appears opposite was made in furtherance of previous studies of this region in attempts to trace the course of the supposed extension of the Hypatia rille towards Censorinus B – so boldly depicted on some maps. Whilst a doubtful feature following a fragmentary course eastwards in continuation of the general direction of the Hypatia rille has been traced, nothing has been seen by the writer which at all resembles the very distinct rille shown on certain charts.

The observation of 1987 Sept.12/13 shows darkening bands running along the site of what appears, under opposite illumination, to be a fault-line or V-shaped 'railway cutting'. In the present drawing the strong impression is conveyed that we are viewing the southern-facing slope of this 'cutting' which is very obliquely and dimly lit by the setting Sun. The darkening slope is interrupted midway by the broad, uneven ridge which runs north from the principal crater (Torricelli B).

It seems at least possible that earlier observers were misled by the linear shading or, in reverse light, brightening of these slopes giving the appearance of continuity with the main Hypatia rille – particularly since the 'strike' has a similar, if not identical, direction.

Viewing the matter speculatively, the slopes may well indicate the vestigial remains of an associated and collapsed fault running along the lines of the original rille, but obviously more observational work at the appropriate times is indicated.

Added note The 4 mile crater Torricelli B has been kept under constant surveillance by members of the B.A.A. Transient Lunar Phenomena Dept since 1983 January 29 when possibly the most significant and well-authenticated TLP event of recent times occurred. A continual watch is maintained for the recurrence of further outbursts, not only here but in other suspect areas. This is not the place to discuss the validity of results achieved to date in this particular field of lunar investigation, but these findings will doubtless receive, in due course, a thorough analysis from an authoritative quarter.

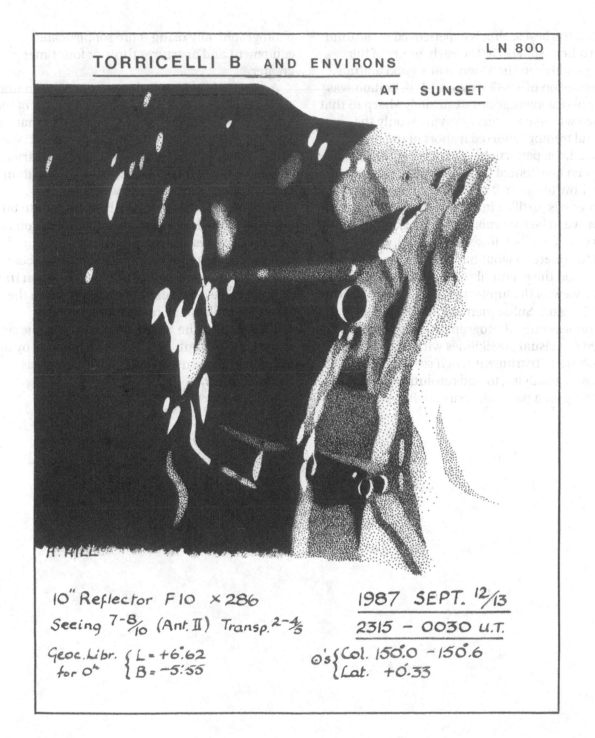

TORRICELLI B AND ENVIRONS

AT SUNSET

LN 800

H. HILL

10" Reflector F10 ×286

Seeing 7-8/10 (Ant. II) Transp. 2-4/5

Geoc. Libr. { L = +6°.62
for 0ʰ { B = -5°.55

1987 SEPT. 12/13

2315 - 0030 U.T.

⊙'s { Col. 150°.0 - 150°.6
{ Lat. +0°.33

Torricelli and environs

Very often the best seeing is experienced in autumn (northern hemisphere) in the early hours of the morning and when the Moon is at a good altitude. On the occasion of 1987 August 14, definition was not only above average but exquisitely sharp so that the image was like a steel engraving – only the occasional tremor rendered it short of the magic Antoniadi I, i.e. perfection. A very real problem arises when confronted with such an opportunity, namely, how to utilise it to fullest advantage and whether one is justified in embarking upon what might prove to be too ambitious a programme in the likely event of a fall-off in seeing quality. It was decided to adhere to what had been planned in advance, and the eventual result was possibly my best-ever view of the lunar surface – certainly of the Torricelli region. Subsequent comparison with quality professional photography of the same area indicated the visual possibilities with only a moderate-sized instrument when seeing is first-class. It was gratifying to find resolution of the image being on a par with results achieved photographically using more sophisticated equipment and apertures three or four times greater.

An example of how little we can see, even under such conditions, of the small-scale features of the lunar surface is demonstrated by the fact that the smallest crater on the drawing – distinguishable as such – is still $\frac{1}{2}$ mile across, which by terrestrial standards is still a very large object. In less than perfect seeing, such a crater would not be recognisable so that if we are bold enough to take a sternly realistic position, this means that on most nights our views and attempted portrayals of the lunar scene, impressive though they may seem, are mere caricatures at best of the reality. When those unpredictable occasions come along, when the observer thrills to the realisation of perfect definition, then the Moon is a truly awesome sight to the beholder but, alas, it is accompanied by an overwhelming sense of impotence as regards representation!

TORRICELLI and Environs
— sunset.

L.N. 799

A

2 very bright peaks

B

Scale of drawing:
175" to lunar diameter

1987 August 14
0220 - 0305 U.T.

⊙'s Colong. 145°.2 - 145°.6
10" Newtonian Reflector F/10 ×286
Seeing 8-9/10 (Ant. I - II) Transp: good.

Note.
Exquisite definition. and my best ever view of the lunar surface in detail —

H. Hill.

The last stages of illumination on Geber, Almanon and Abulfeda

One of those unanticipated sights that it is possible to catch at the terminator and remarkable enough to compel a representation.

It will be seen that the crests of the two rings Geber and Almanon have broken up into sections with the lower two-thirds of the latter formation obscured by shadow from the elevations to the west. The southern limit of this shadow is well defined. The remnants of Almanon appear in the drawing as at commencement of the observation but the two easternmost fragments of wall had gone 35 min later and the remaining components had dulled appreciably. The rapidity with which features appear/disappear (as the case may be) means that the process of making a half-hour drawing cannot give a precise account of the state of lighting which a photograph can provide, but there are often compensations in the visual record in the minuteness of detail captured in moments of the greatest steadiness which, all too often, the camera fails to record.

It needs little imagination to visualise this same scene from a point in space, say, 50 miles above the surface with these enormous fragmented rings glowing in the last rays of sunlight before being swallowed up into the lunar night. These illuminated coronets would appear much more detailed than a telescopic account from 800 miles away could afford! A truly magnificent spectacle of stark contrasts and utterly desolate grandeur unlike any earthly scene.

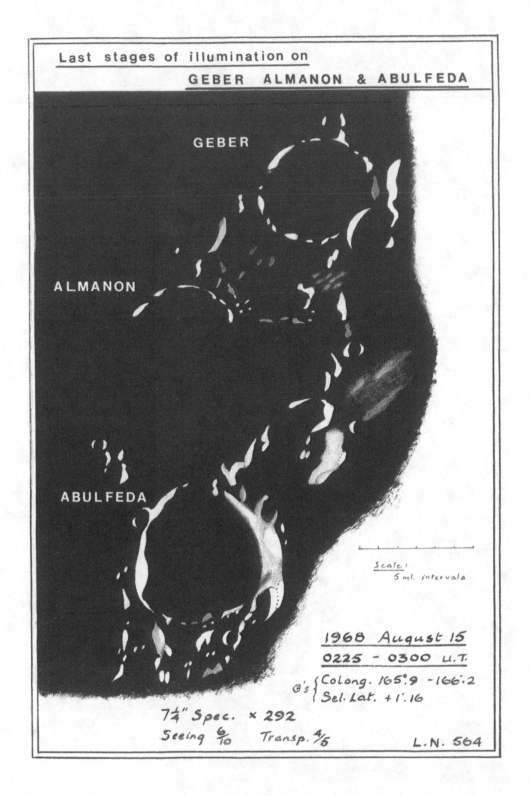

Last stages of illumination on
GEBER ALMANON & ABULFEDA

GEBER

ALMANON

ABULFEDA

Scale:
5 ml. intervals

1968 August 15
0225 – 0300 U.T.
⊕'s { Colong. 165°.9 – 166°.2
{ Sel. Lat. + 1°.16

7¼" Spec. × 292
Seeing 6/10   Transp. 4/5

L.N. 564

Quadrant IV – Section 14

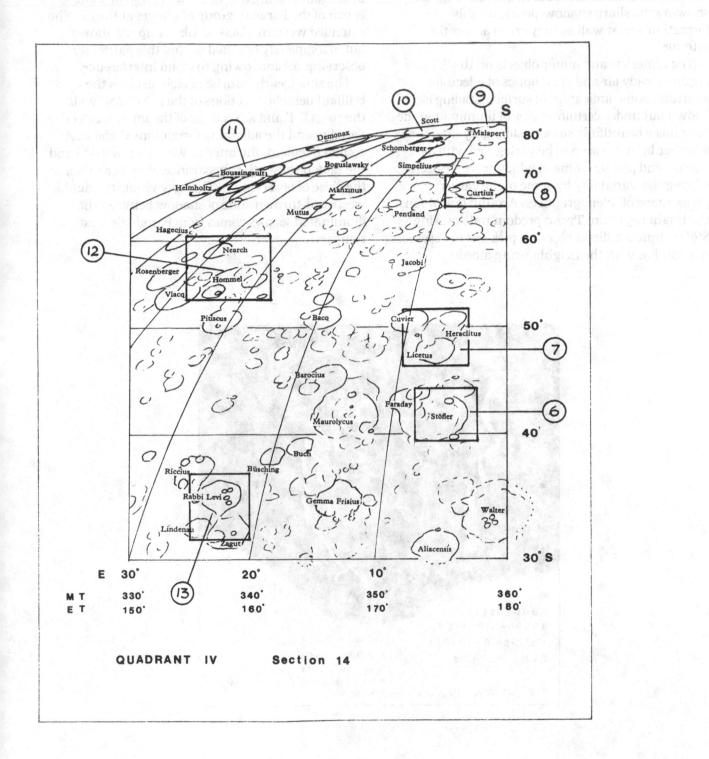

QUADRANT IV   Section 14

Stöfler

There are few more dramatic scenes than the long shadows cast by a low Sun across the floors of the larger walled-plains such as Stöfler, and the sunset drawing was made at probably the best stage for showing the sharp shadow-profiles of this formation's west wall as they creep across the interior.

The craterlets and minor objects on the floor require steady air and good optics of adequate aperture. Some indication of surface shading is shown but under certain angles of illumination the floor has a beautifully smooth steel-grey hue; under a higher light it is seen to be variegated with dark streaks and patches some of which were suspected of irregular variability by some of the earlier generations of selenographers. Around Full Moon the bright rays from Tycho predominate making Stöfler quite a difficult object to pick out to observers not familiar with the neighbouring albedo 'landmarks' which help as guides to the location of Stöfler.

Stöfler is some 69 × 85 miles across, its deformation being due to the incursion at some epoch of the Faraday group of craters at the SE. The extended western glacis of this group are shown, but incompletely depicted during this particular observing session, owing to cloud interference.

The small, early sunrise sketch displays the brilliant detached sections of the great west wall, the rings E, F and K, sections of the inner walls of Faraday and the adjoining formations C and P. At commencement, the interior was shadow-filled and the sketch shows the appearance of Stöfler's floor at the close of observation. The now visible triangular westward-thrown cone of shadow between the illuminated sectors comes from Faraday's west rampart.

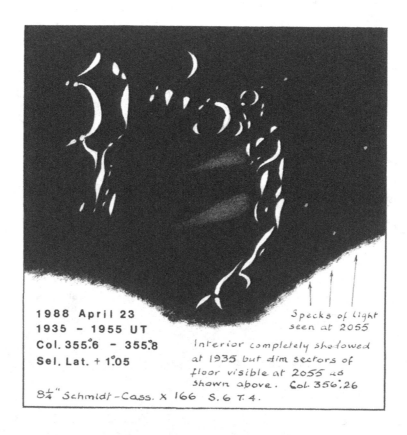

1988 April 23
1935 – 1955 UT
Col. 355°.6 – 355°.8
Sel. Lat. + 1°.05

Specks of light seen at 2055

Interior completely shadowed at 1935 but dim sectors of floor visible at 2055 as shown above. Col. 356°.26

8¼" Schmidt-Cass. × 166 S.6 T.4.

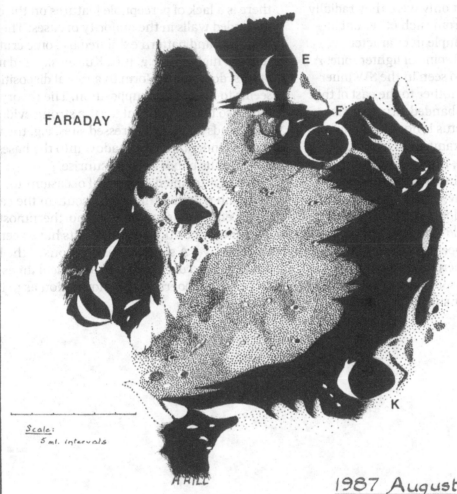

The Floor of STÖFLER at sunset. LN 799

FARADAY

E

F

N

K

Scale:
5 ml. intervals

A HILL

10" Newtonian Reflector
F/10 ×286
Seeing 6-7/10 (Ant. II) Transp. 4/5

1987 August 16
0115 - 0205 U.T.
⊙'s { Colong 169°.1 -169°.5
       Sel. Lat. +1°.01

The Licetus–Heraclitus Region

Under suitable conditions this region is seen to abound with interesting features, particularly along the inner slopes and central ridge of the irregular enclosure Heraclitus, but at the time of this observation my attention was drawn to five or six well-marked bands on the western and southern inner slopes of Licetus. Not only were they radially disposed and equidistant from each other but at least two were seen to be duple in character by reason of a narrow central spine of lighter tone. A broad forked band was also seen in the SW inner wall of Heraclitus D and another to the east of this.

The presence of distinct banded structures in formations as large as Licetus is not very common although I have noticed examples in the twin craters Steinheil and Watt, Rheita and elsewhere. The comparatively gentle inner slopes of Licetus rather give the lie to the theory that, because most bands are to be found on slopes equal to, or in excess of, the natural 'angle of repose', it necessarily follows that the banded appearance must be due to a landslide mechanism whereby the lighter surface 'soil' has, for whatever reason, slipped down the incline to reveal the darker subsurface material.

This idea had some support when core tubes returned from the Apollo missions showed that the lunar regolith comprises layers of both light and dark material. Some investigators, however, have maintained that bands are the result of lava flows originating from vents or craterlets at their head but there is a lack of perceptible features on the crests of banded walls in the majority of cases. The fact that the band pattern exhibited by some craters under a high Sun, e.g. Birt, Kunowsky and many others, does not conform to a radial disposition also seems to negate this supposition. The theory of lava flow also fails to account for the strong evidence that bands lie along depressed sites, e.g. the well marked indentation of shadow into the bases of the Aristarchus bands at early sunrise.

I have attempted on several occasions to ascertain whether this effect occurs in the case of the Licetus bands but only the northernmost bands are suitably positioned and results have been inconclusive. A series of observations of the bands made under a declining Sun at critical times might give useful topographical information as to their contours.

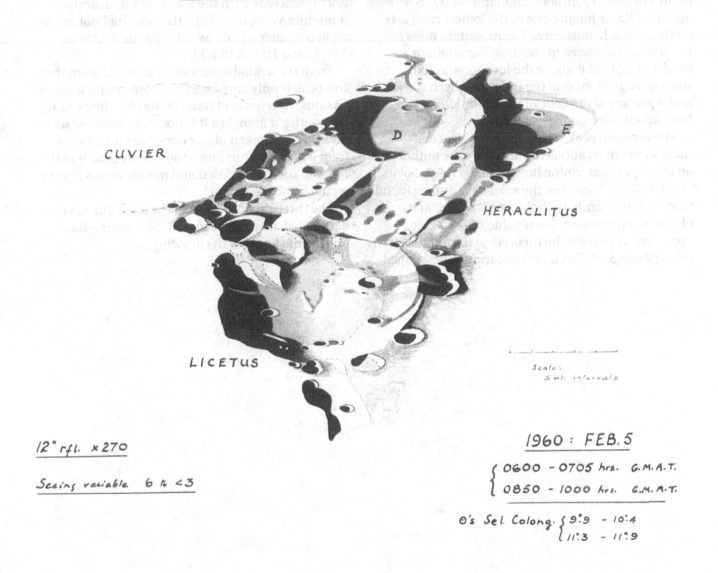

L.N. 459 — THE LICETUS-HERACLITUS REGION —

CUVIER

D

E

HERACLITUS

LICETUS

V

Scale:
5 ml. intervals

12" rfl. × 270

Seeing variable 6 to <3

1960 : FEB. 5

{ 0600 – 0705 hrs. G.M.A.T.
0850 – 1000 hrs. G.M.A.T.

O's Sel. Colong. { 9°.9 – 10°.4
11°.3 – 11°.9

Curtius

A grand object, some 58 miles in diameter, situated in the congested southern highlands at 67° S 4° E and notable for having one of the loftiest ramparts on the visible hemisphere. The mountain mass on the NW and arrowed in the drawing attains a height of 22 000 ft above the interior and the shadow engulfs most of the inner northern slopes and a portion of the floor in the lunar afternoon from about colongitude 160° onwards.

The arrowed peak has been the subject of a number of observations at sunrise by the author in an attempt to determine how soon, by reason of its great altitude, it catches the solar rays and he found that on 1987 Jan.6, with the Sun's sel.lat. at its almost maximum southern value of $-1°53$ it appeared far beyond the theoretical terminator at colongitude 349°7 as a brilliant triangular-shaped mass accompanied by two lesser heights to its north; whereas with the Sun's sel.lat. near its minimum value of $+1°45$ the peak had not made an appearance as late as colongitude 352°9 on 1989 June 10 (2235 UT).

From the available measures it would seem that this peak is only surpassed by those on the walls of Casatus, Newton and some of the elevations along the southern limb, but it is not clear whether such measures are taken above depressed interiors or from a mean datum line – the establishment of the latter in such a confused and mountainous region would be no mean feat!

The three satellite rings to the SE of Curtius are designated after the formation Simpelius – itself outside the limits of the drawing.

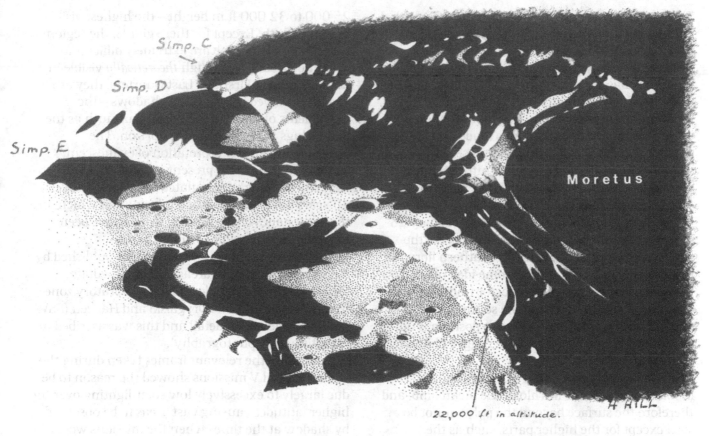

Simp. C

Simp. D

Simp. E

Moretus

22,000 ft in altitude.

H HILL

CURTIUS and environs at sunrise – LN 792

8¼" Schmidt.-Cass. ×245

Seeing 6-7/10 [Ant. II]

Transp. 5/5

Geoc. Libr { L = +7°08
for 1730 { B = -1°.11

1987 January 7

1645 - 1735   then 1840 - 1900

⊙'s { Col. 1°.78 - 2°.15   2°.78 - 2°.90
{ Sel. Lat. -1°.54

The South Polar Regions

The whole region contained within the 80° south parallel and as far over on to the averted side as libratory conditions will permit, came under systematic survey in the early 1950s by the author. This work has continued up to the present, and has accounted for the greater part of the time he has spent in lunar observation. This investigation now comprises hundreds of drawings made under every possible condition of lighting and presentation and will form the subject of a detailed future treatise, consequently this portfolio is not the place to discuss topographical results.

However, the observation of 1967 February 23 is included to show a general view of part of the region as seen under favourable southern libration just before Full Moon. It gives some idea of what is possible using only a small reflector under suitable conditions. On this drawing the south pole is marked with a cross but its precise position is not easy to determine because of the rugged character of the whole area.

It will be appreciated that at, or around, Full Moon the terminator lies along the mean limb, and therefore the surface beyond the pole cannot be seen except for the higher parts, such as the mountains M4, M5 and M6 (the original designations given to these elevations by E. A. Whitaker and which have been retained). These actually lie on the averted side and range from 25 000 to 32 000 ft in height – the highest visible from the Earth. Except for these giants, the regions beyond the mean limb are extremely difficult to examine, because although *theoretically* visible in their entirety at First and Last Quarters, they are heavily involved in mountain shadows – the movements of which are chiefly in azimuth as the Sun swings around the lunar horizon. The identification and interpretation of evanescent features such as these present great problems. Nevertheless, the accumulated observations over 40 years have yielded much information which, in the early stages of the project, would have been regarded as impossible to obtain.

The stereographic projection charts published by NASA show extensive areas in the immediate vicinity of the south pole and in the libratory zone beyond the formations Drygalski and Hausen (SSW limb) to be devoid of detail and this was ascribed to 'unsatisfactory photography'.

Analysis of the relevant frames taken during the Orbiter IV and V missions showed the reason to be due largely to excessively low solar lighting over the higher latitudes causing vast areas to be obscured by shadow at the times when the missions were undertaken. Resumption of direct visual studies, after what had appeared during the photographic surveys to be a lost cause was, therefore, justified!

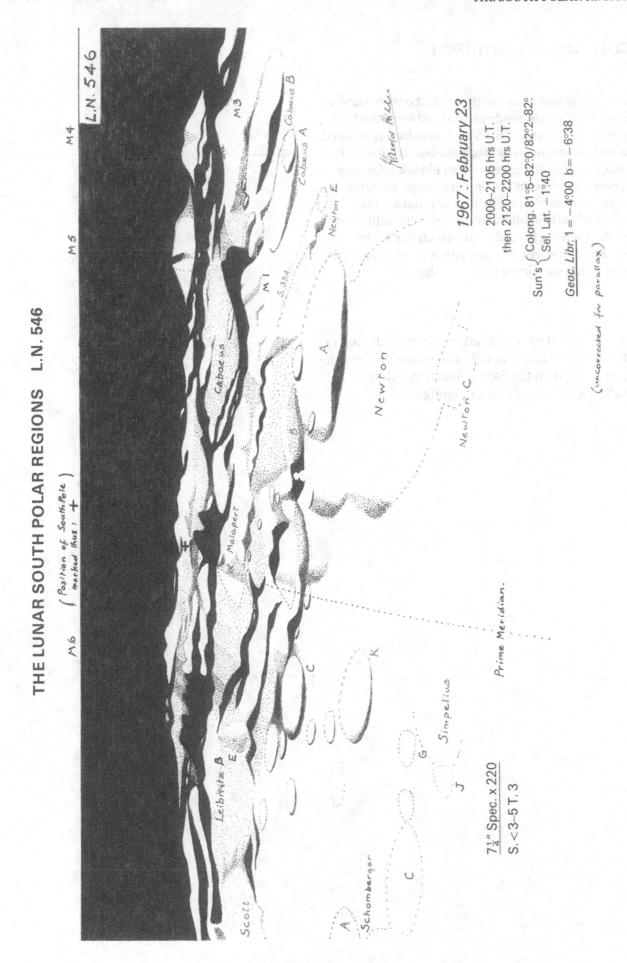

THE LUNAR SOUTH POLAR REGIONS   L.N. 546

M6   (Position of South Pole marked thus: +)

M5   M4.

L.N. 546

M3

Cabaeus B

Cabaeus A

Newton E

S.354.

M1

Cabaeus

A.

B.

Newton

Newton C

Malapert

C

K

Leibnitz-β

E

Simpelius

J

G

Scott

C

Schomberger

A

Prime Meridian.

7¼" Spec. x220
S. <3–5 T.3

1967: February 23

2000–2105 hrs U.T.
then 2120–2200 hrs U.T.

Sun's { Colong. 81°.5–82°.0/82°.2–82°.
        Sel. Lat. –1°.40

Geoc. Libr. 1 = –4°.00 b = –6°.38

(uncorrected for parallax)

189

Scott and Amundsen

Two early observations of these most appropriately named formations, made at, and just after the First Quarter phase. At 85° S 85° E Amundsen is situated just within the mean hemisphere but its interior is difficult to examine except when libration favours its presentation as in the two drawings. Effective viewing is limited to morning illumination only because of early shadow under opposite light from the lofty Liebnitz Beta plateau which lies to the west. Further north, the formation Scott is similarly affected but to a somewhat lesser degree.

Note

At the time of these observations Greenwich Mean Astronomical Time was still in operation where O h began at noon on the date in question – thus avoiding a change of date at midnight.

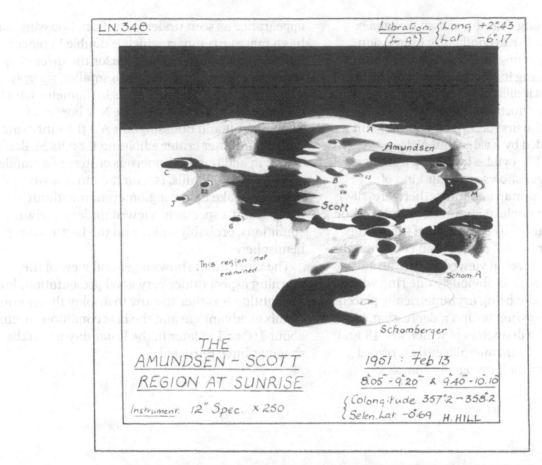

LN. 348

Libration. {Long +2°.43
(for 9ʰ) {Lat −6°.17

THE
AMUNDSEN − SCOTT
REGION AT SUNRISE

Instrument. 12" Spec. × 250

1951 : Feb 13

8.05ᵐ − 9.20ᵐ & 9.40 − 10.10

{Colongitude 357°.2 − 358°.2
{Selen. Lat. −0.69   H. HILL

LN. 348

Libration {Long −0°.10
(for 9ʰ) {Lat −6°.81

THE
AMUNDSEN − SCOTT
REGION − MORNING

Instrument. 12" Spec. × 250

1951 : Feb. 15

8.00ᵐ − 9.30ᵐ & 9.50 − 10.50

{Colongitude 21°.5 − 23°
{Selen. Lat. −0°.66   H. HILL

Boussingault

It is hardly surprising that this massive complex should have given rise to conflicting impressions among the early selenographers – a contributory factor, perhaps, being its position towards the SE limb which makes it difficult to examine satisfactorily under normal conditions.

It is not clear who first described Boussingault as an enclosure divided by a ridge, but Neison compared it with Phocylides in having a similar cross-wall and Elger shows a similar kind of configuration in his map and must, therefore, have considered it to be duple in character. As late as the 1930s it was regarded as consisting of three rings within each other – offset rather than concentric in arrangement. However, if viewed under the most favourable libration, it is obviously one ring within another – the smaller being asymmetrically placed to the NW – a depression within a depression, in fact. Both rings, the diameters of which are 48 and 82 miles, are deeply and magnificently terraced which accounts for their extremely complex appearance as seen under a low Sun. Boussingault has a massive rampart which is double in places and may have been responsible for the three-ring concept of the whole. There is a smaller, sharply defined crater K, some 19 miles in diameter which occupies the space between the NW border of Boussingault and Boussingault A – the inner ring with yet another crater adjoining K on its SE flank. These, in addition to numerous craters of a smaller order, ridges and hills, etc, on the inner walls and interiors, make up a conglomeration without parallel and a spectacle, viewed under the right conditions, probably unique on the Earth-facing hemisphere.

The illustration shows a general view of the morning aspect under very good presentation, but the lighting is rather too low to display the interior details to advantage and the best conditions occur about 10° or 15° later in the lunar day when the shadows have retreated.

BOUSSINGAULT

and neighbouring formations

under morning illumination

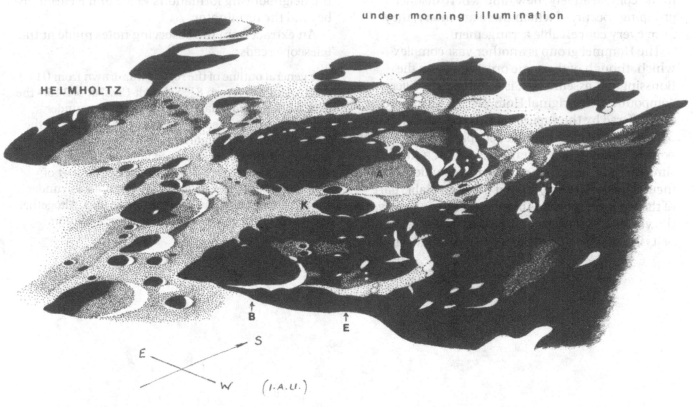

Limb

HELMHOLTZ

A

K

B

E

E — S

W — (I.A.U.)

Librational values very favourable
for this particular limb
$\begin{cases} L = +7°.45 \\ B = -6°.09 \end{cases}$

$8\frac{1}{4}''$ Schmidt-Cass. ×245   Seeing $\frac{5-6}{10}$ Transp. $4-2\frac{2}{5}$

<u>1989 : April 10</u>
<u>1920 - 2040 U.T.</u>

⊙'s $\begin{cases} \text{Colong. } 326°.5 - 327°.2 \\ \text{Sel. Lat. } +1°.14 \end{cases}$

The Hommel Group

Of all the quadrants, the fourth par excellence, displays the lunar surface broken and rugged beyond all description, presenting every type of formation in abundance from the obviously very old to the comparatively 'new' and where distinct groupings occur, of which there are many, they form every conceivable arrangement.

The Hommel group is another vast complex which, though of the same order of size as the Boussingault assemblage, is a completely different compound. The original Hommel, 75 miles across, has suffered by the superimposition of ring A, some 35 miles in diameter which occupies a considerable portion of the interior on the NE, whilst a ring of similar size, C, together with D, have made incursions on the SW, so that an appreciable part of the old formation has been obliterated. However, the vast SE rampart remains intact and the breadth of its inner slope and the wealth of detail it contains gives some conception of what the original Hommel must have been like.

The drawing is an attempt to portray its present appearance as seen under late evening light with the neighbouring formations Vlacq and Nearch just beyond the terminator.

An extract from the observing notes made at the telescope reads:

The general outline of the region was drawn from 0110–0220 using the driven $8\frac{1}{4}''$ Schmidt-Cass. at $\times 245$ – the details being filled in later with the 10'' Spec. under improved seeing. There was a mass of fine detail on the inner SE slopes of Hommel which could not be properly grasped under the prevailing conditions – an obvious subject for some future occasion. The SE rampart of Hommel H seemed extraordinarily 'hummocky' under the present lighting and must be looked into. Altogether, the whole district is one deserving closer attention.

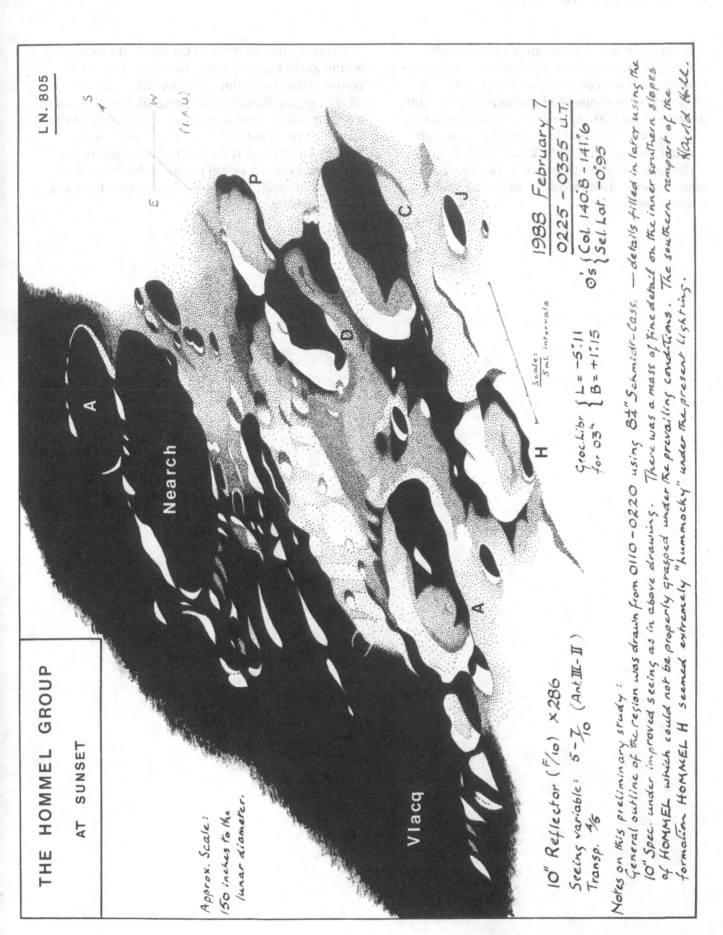

Rabbi Levi and Zagut

Although the fourth quadrant is crowded with walled-plains, ring-plains and craters of every size and type, it is not commonplace to find discrete groupings of medium-sized contiguous craters on their floors such are to be found within Rabbi Levi. This is a quite remarkable assemblage although another such group is to be found in the larger walled-plain Walter (also in Sect.14). The latter makes an equally interesting subject for study under a low Sun.

Mention must be made of the bright linear feature within Zagut running from the crater A to the SE border of that formation. A short section can be seen emerging from the sunrise shadow – obviously a continuation – and suggesting that it is either the face of an eastward-looking fault or the raised border of a rille; yet it has not been possible to find such a marking on any of the available maps and attempts at verification from photographs have not been successful.

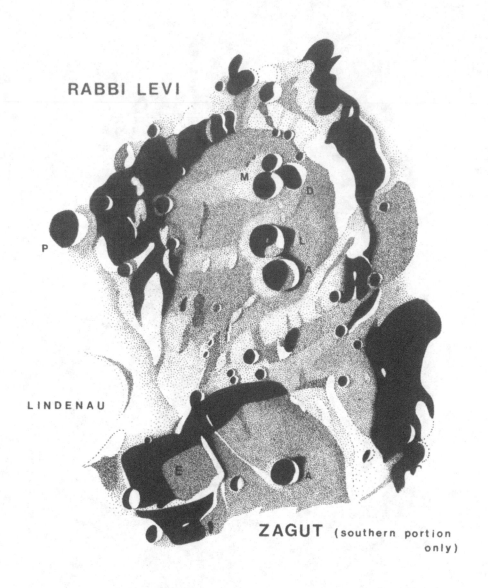

RABBI LEVI

LINDENAU

ZAGUT (southern portion only)

10" Reflector (Newt.) × 286

Seeing: 5 - 7/10

Transp: 4/5

Geoc Libr   L = -7°24
for 2030   B = -6°61

1989 March 13

2000 - 2220 U.T.

☉'s { Colong. 345°6 - 346°8
      { Sel. Lat. + 0°50

Quadrant IV – Section 15

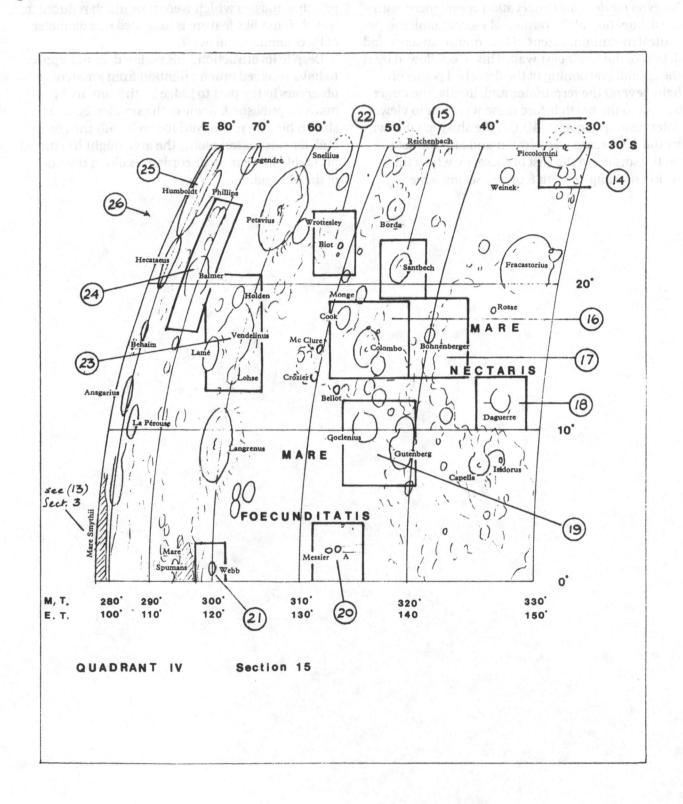

QUADRANT IV   Section 15

Piccolomini and the Altai Mts

This region overlaps two of the section charts, namely 13 and 15.

Notes on the drawing Observation commenced with the delineation of the outline of Piccolomini, the central mountain, extent of the interior shadow and details of the inner east wall. This was followed by the careful positioning of the detached points of light beyond the terminator and, finally, the larger masses to the north before these were lost to view. After resumption at 0340 UT. the shadows thrown by the Altai scarp were drawn and also the detail on the surface to the east of these, by which time seeing had improved to 8 on my seeing scale –

equivalent to Antoniadi II. By 0430 (colongitude 146°) all the smaller detached points of light to the E and NE of Piccolomini had gone except the principal masses which were now much reduced in size. A dome-like feature is indicated one diameter of Piccolomini to the west.

Despite its attraction, this region does not appear to have received much attention from amateur observers in the past to judge by the paucity of material published. Some of the smaller features shown here do not accord too well with the charts and, for this reason alone, the area might be studied profitably to clear up discordances along the lines of the old tradition.

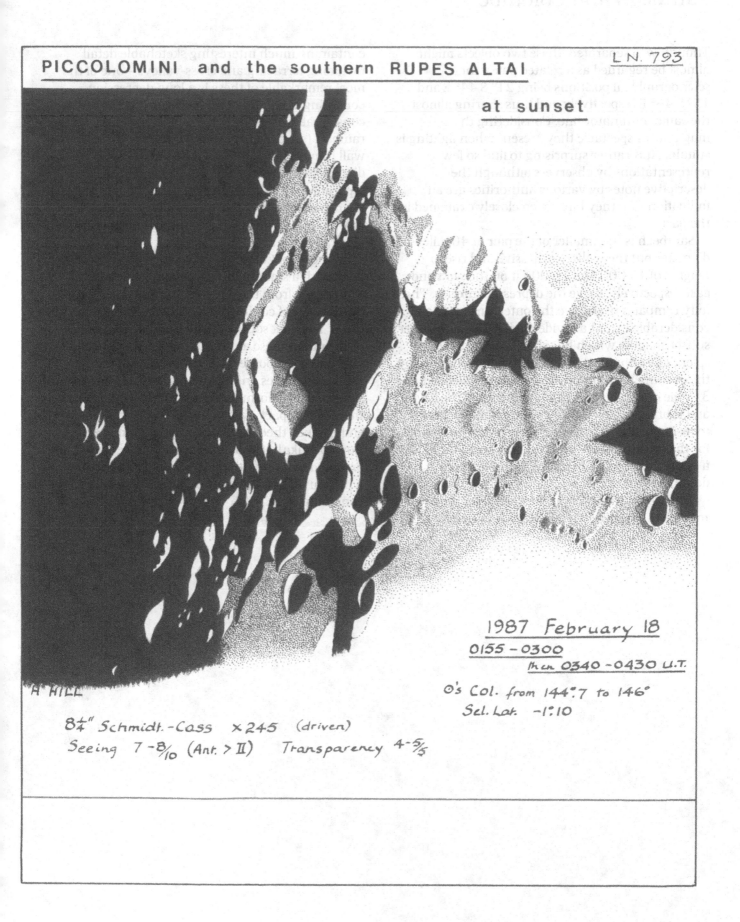

PICCOLOMINI and the southern RUPES ALTAI

L.N. 793

at sunset

1987 February 18
0155 - 0300
then 0340 - 0430 U.T.

⊙'s Col. from 144°.7 to 146°
Sel. Lat. -1°.10

A HILL

8¼" Schmidt.-Cass ×245 (driven)
Seeing 7 -8/10 (Ant. > II)   Transparency 4-5/5

Santbech and Colombo

Though well separated, these two objects might almost be regarded as a related pair – their selenographical positions being 21° S 44° E and 15° S 46° E respectively and thus sharing almost the same terminator line. Considering the magnificent spectacle they present when lighting is suitable, it is rather surprising to find so few representations by observers although the descriptive notes by various authorities are an indication that they have been closely examined in the past.

Santbech is the smaller of the pair at 40 miles diameter but the walls are massive and rise to heights of 15 000 and 10 000 ft on the west and east respectively above the depressed interior. These lofty ramparts rise above the outer surface to a considerable extent as evidenced by the large outer sunset shadow which has sharp peak profiles splendidly enhanced just prior to being engulfed by the evening terminator. A very sharply defined $3\frac{1}{2}$ mile crater lies just below the crest of the E wall and is brilliant at this stage. The detached marking to its immediate east would seem to be responsible for the principle spire marking this position. This is a region abounding in interest but Santbech dominates the scene.

Colombo is some 49 miles in diameter with massively structured walls 8000 ft in height containing much interesting sketchable detail including terracing and cross-valleys – one of the most remarkable of these is a long depression consisting of confluent craters which occupies a considerable portion of the inner slope of the eastern rampart (see drawing). The continuity of the west wall is interrupted by the intrusion of the 25 mile diameter ring Colombo A which is, in itself, a separate subject for study. On the floor of Colombo is a central group of five or six hills arranged in an incomplete circle but suggesting the remnants of an old crater-ring.

To the SE of Colombo is the low-walled enclosure Cook with a smooth dark floor almost devoid of detail apart from the $3\frac{1}{2}$–4 mile crater A. A sharply defined pair of confluent craters breaks the crest of the W wall just south of centre and in the process of formation these appear to have demolished the wall down to floor level. Curiously, as in many similar instances on the Moon, there are no obvious signs of wall deformation immediately adjacent to the site of impact – a point which advocates of endogenous theories of crater formation are not slow to make! Another confluent pair seem to have broken down or bestrode the N wall in a manner which invites closer scrutiny under better conditions.

SANTBECH

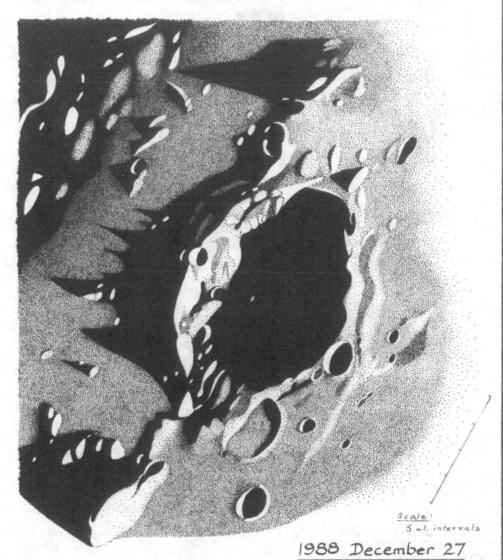

Scale:
5 ml. intervals

1988 December 27

0305 - 0420 U.T.

10" Reflector (Newtonian) ×286
with W8 filter.
Seeing 7-8/10 at commencement
then slowly deteriorating over the
observing period to 4/10
Transparency variable 3-4/5

⊙'s { Colong. 132°2 - 132°9
     { Sel. Lat. -1°37

LN 816

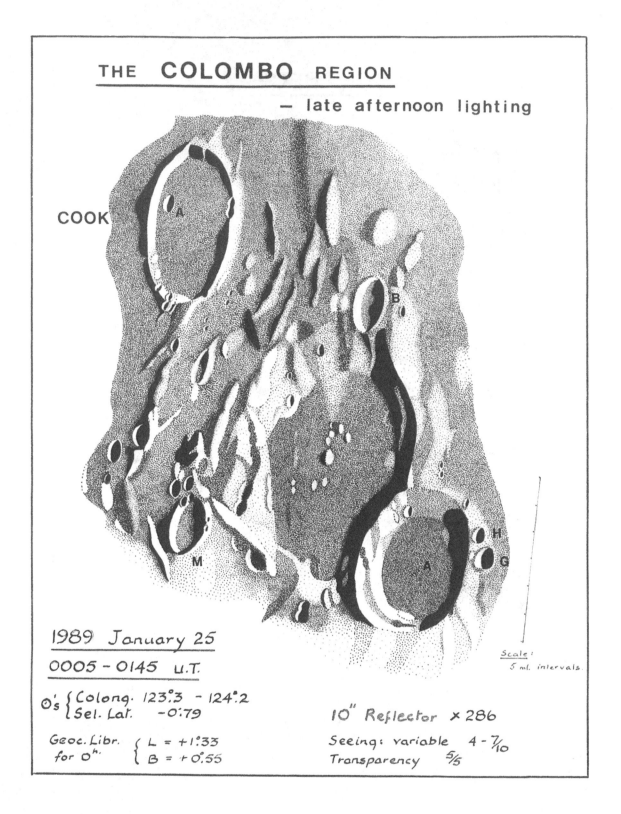

THE **COLOMBO** REGION

— late afternoon lighting

COOK

A

B

H

G

A

M

1989 January 25

0005 – 0145 U.T.

$\odot$'s { Colong. 123°3 – 124°2
{ Sel. Lat. –0°79

Geoc. Libr. { L = +1°33
for 0ʰ.   { B = +0°55

10" Reflector × 286

Seeing: variable 4 – 7/10
Transparency 5/5

Scale:
5 mi. intervals.

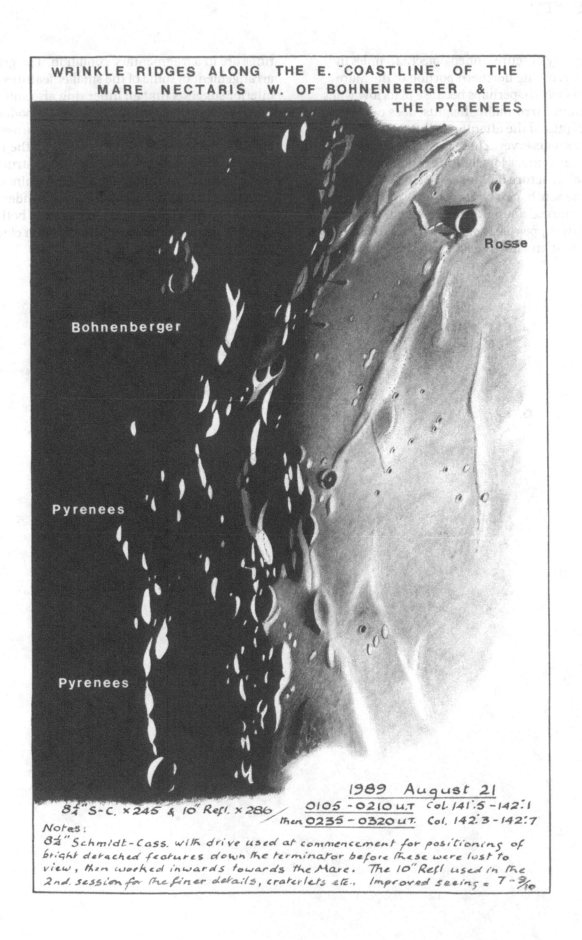

WRINKLE RIDGES ALONG THE E. "COASTLINE" OF THE MARE NECTARIS W. OF BOHNENBERGER & THE PYRENEES

Rosse

Bohnenberger

Pyrenees

Pyrenees

1989 August 21

8¼" S-C. ×245 & 10" Refl. ×286 / 0105 - 0210 u.т Col. 141.5 - 142.1
then 0235 - 0320 u.т. Col. 142.3 - 142.7

Notes:
8¼" Schmidt-Cass. with drive used at commencement for positioning of
bright detached features down the terminator before these were lost to
view, then worked inwards towards the Mare. The 10" Refl used in the
2nd. session for the finer details, craterlets etc. Improved seeing = 7 - 9/10

Daguerre

A relatively obscure ruined object, 28 miles in diameter, on the northern portion of the Mare Nectaris which, perhaps because of its low walls and minor surrounding heights, does not appear to have captured the attention of lunar observers as much as it deserves. There are few drawings extant. When seen under low lighting, however, the original structure of Daguerre before its inundation from the south, can be easily made out. That it had a well-marked double rampart is evident, but, whilst the upper parts of these are relatively intact to the west and north-east, large sections of both rings are in a fragmentary condition. The general arrangement of some of the smaller features on the interior suggests that an inner ring also suffered from the destructive agency of marial flooding.

Daguerre is seen to best effect under sunset lighting, indeed, a drawing sequence of the very last stages of illumination would form an instructive topographical study but, like so many ruined formations, it should also be observed under a high Sun, when much of the sub-structure of both partially ruined and 'ghost' rings is often clearly revealed.

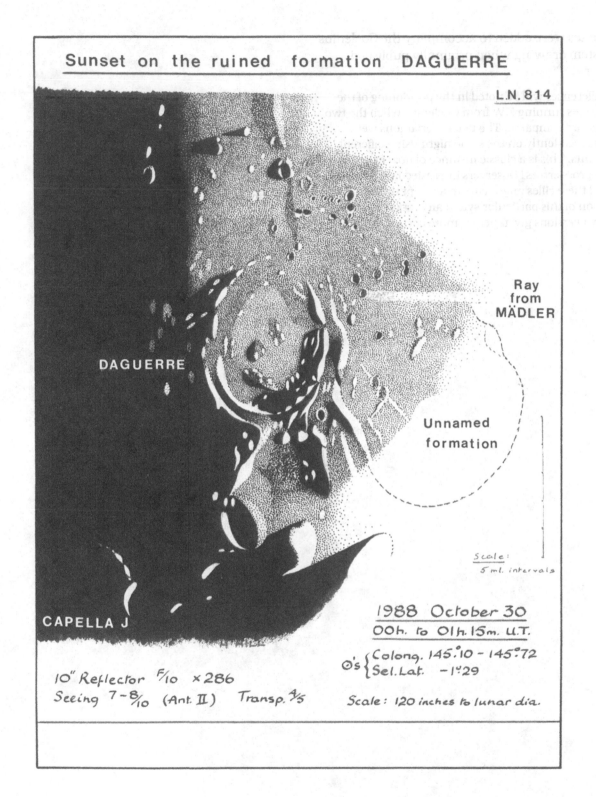

Sunset on the ruined formation DAGUERRE

L.N. 814

Ray from MÄDLER

DAGUERRE

Unnamed formation

Scale: 5 ml. intervals

CAPELLA J

1988 October 30
00h. to 01h. 15m. U.T.

⊙'s { Colong. 145°.10 - 145°.72
      { Sel. Lat. -1°.29

10" Reflector F/10 ×286
Seeing 7-8/10 (Ant. II) Transp. 4/5

Scale: 120 inches to lunar dia.

The Goclenius Rille System

Some notes were added to accompany the Goclenius rille system drawing when originally published elsewhere:

Some differences will be noted in the positioning of the parallel rilles running NW from Goclenius when the two drawings are compared. The two observations were made independently on the same night using different instruments. This is a classic instance of how easy it is to make gross errors! Observers interested in tracing the course of these rilles might care to attempt their own impression of this particular sytem and establish which of the two versions given here is more correct.

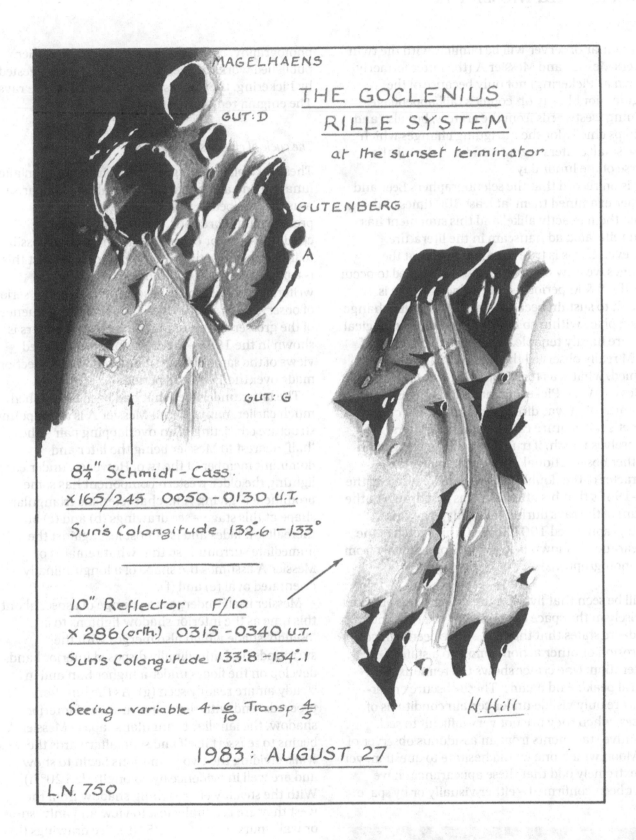

MAGELHAENS

GUT: D

THE GOCLENIUS
RILLE SYSTEM

at the sunset terminator

GUTENBERG

A

GUT: G

8¼" Schmidt - Cass.
×165/245 0050 - 0130 U.T.

Sun's Colongitude 132°.6 - 133°

10" Reflector   F/10
×286 (orth.) 0315 - 0340 U.T.

Sun's Colongitude 133°.8 - 134°.1

Seeing - variable 4 - 6/10 Transp 4/5

H. Hill

1983 : AUGUST 27

L.N. 750

Messier and Messier A

Every lunar observer will be familiar with the twin craters Messier and Messier A (the latter formerly known as Pickering) not only because of the singular double-ray or 'comet's tail' appendage running westwards from the craters, but also and perhaps chiefly for the intriguing changes which these small craters seem to undergo during the course of the lunar day.

It is on record that the selenographers Beer and Madler examined them 'at least 300 times and found them exactly alike' and this statement has been reiterated ad nauseam in the literature. However, if this is true then it means that the changes we now observe must have started to occur *after* the B & M period – an argument which is difficult to sustain because claims of secular change taking place within so brief a period of selenological time are hardly tenable. It has also to be asked: if B & M really observed the Messiers as often as claimed, what was their object in so doing? Professor W. H. Pickering made the suggestion that their attention was directed mainly to the curious 'comet's tail' feature rather than the craters themselves which, if true, has some bearing upon another observational problem, namely, the character of the double rays. Goodacre wrote in the mid-1930s that his attention was first drawn to the nature of the rays during the study of a Paris photograph dated 1901 Sept. 30 (his sketch of the appearance is shown below and is presumably from the photograph).

It will be seen that five obscure crater-rings are sited precisely in the space contained by the rays and Goodacre states that these '. . . have been reduced by erosion or other action almost to a state of obliteration. One crater shows the remains of a central peak'. And again: 'These obscure crater-rings are only visible under certain conditions of phases, when they are not very difficult to see'. Definitive statements from an assiduous observer of the Moon which one would hesitate to question, yet it is extremely odd that these appearances have never been confirmed – either visually or by space

photography. Nor is there mention in the earlier published works of Beer and Madler if, as suggested by Pickering, they paid special attention to the rays. The enigma remains unresolved.

The cycle of changes

These take place with great regularity lunation after lunation and any apparent anomalies which arise must be ascribed to the effects of physical presentation (libration), the vagaries of seeing conditions and/or observational errors due possibly to inherent subjectivity in the observer. Whilst this is a purely personal opinion on the part of the writer, it is one which is based upon a lengthy series of observations spread over 40 years. The sequence of the grosser changes undergone by the craters is shown in the 1946–48 series: for more detailed views of the same changes I have chosen a selection made over the 1984–88 period.

The latter underline what had been established much earlier, namely, that Messier A is a compound structure consisting of an overlapping pair – the 'half' nearest to Messier being the later and dominant member of the two. However, under early lighting the older western component has some prominence, giving A its characteristic triangular shape at this stage – see drawings (b) and (c) but thereafter it fades and is almost lost against the immediate surrounds so that what remains of Messier A assumes the shape of a longitudinally orientated oval (e) and (f).

Messier itself undergoes a change of aspect about this time as the interior shadow lightens to a 'pseudo' shade which also merges with the surrounds (f). Latitudinally disposed interior bands develop on the floors under a higher Sun and in steady air are readily seen (g). As the lunation proceeds, and with the reappearance of interior shadow, the familiar triangular shape of Messier A begins to reassert itself and soon afterwards the walls dividing the two formations begin to show and are well in evidence by colongitude 120° (i). With the steadily encroaching shadow from the west they are gradually lost to view and only 'stubs' or wall-spurs remain at 125° (k). See drawings (b) to (d) for the reverse effects under morning illumination.

For a more detailed discussion than is possible here, the interested reader is referred to an excellent paper on the subject which appeared in the B.A.A.

Lunar Section magazine *The New Moon*, **3**, No. 1 pp. 13–28 (June 1987). This was written and compiled by R. Moseley, co-ordinator of the topographical department of that Section and it is illustrated by a valuable series of historical drawings by G. & V. Fournier, the two principal observers of the Jarry-Desloges Observatories at Setif and Laghouat in Algeria. The 1913 series of observations they made are of especial interest, not only because of their great detail but direct comparison of simultaneous but completely independent results by the two observers is possible in some instances. They give a fascinating insight into the degree of subjectivity which besets, almost inevitably, the records of even the most well-attested of observers when investigations are conducted at or close to borderline resolution.

The author's observation of 1987 December 9 shows the remarkable spires of shadow which are thrown by the Messier pair as night approaches. Unlike many ray systems, the 'comet's-tail' remains strongly visible under low lighting. An albedo feature bridges the rays at one point – is this a vestige of one of Goodacre's five obscure crater rings? A delicate but well-defined rille with bright eastern border is visible some distance north of the rays.

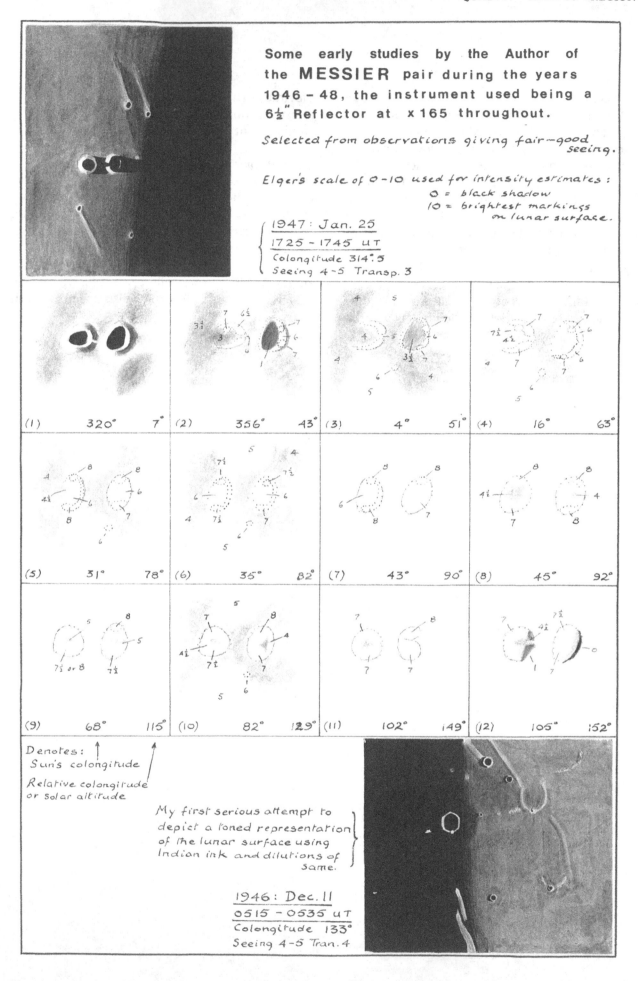

Some early studies by the Author of the **MESSIER** pair during the years 1946 – 48, the instrument used being a 6½" Reflector at × 165 throughout.

Selected from observations giving fair–good seeing.

Elger's scale of 0–10 used for intensity estimates:
0 = black shadow
10 = brightest markings on lunar surface.

1947: Jan. 25
1725 – 1745 UT
Colongitude 314°.5
Seeing 4–5 Transp. 3

(1) 320° 7° (2) 356° 43° (3) 4° 51° (4) 16° 63°

(5) 31° 78° (6) 35° 82° (7) 43° 90° (8) 45° 92°

(9) 68° 115° (10) 82° 129° (11) 102° 149° (12) 105° 152°

Denotes:
Sun's colongitude
Relative colongitude or solar altitude

My first serious attempt to depict a toned representation of the lunar surface using Indian ink and dilutions of same.

1946: Dec. 11
0515 – 0535 UT
Colongitude 133°
Seeing 4–5 Tran. 4

212

MESSIER & A

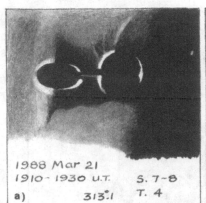

1988 Mar 21
1910 - 1930 U.T.   S. 7-8
a)   313.1   T. 4

1988 May 2
2020 - 2035   S. 6-7
b)   325.7   T. 3

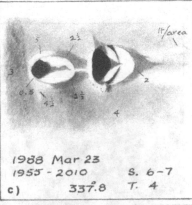

1988 Mar 23
1955 - 2010   S. 6-7
c)   337.8   T. 4

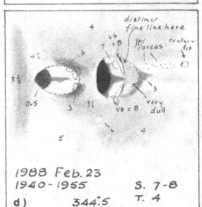

1988 Feb. 23
1940 - 1955   S. 7-8
d)   344.5   T. 4

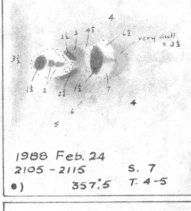

1988 Feb. 24
2105 - 2115   S. 7
e)   357.5   T. 4-5

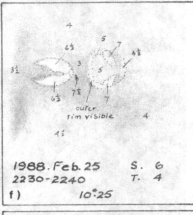

1988. Feb. 25   S. 6
2230 - 2240   T. 4
f)   10.25

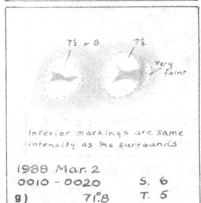

Interior markings are same intensity as the surrounds

1988 Mar. 2
0010 - 0020   S. 6
g)   71.8   T. 5

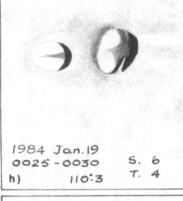

1984 Jan. 19
0025 - 0030   S. 6
h)   110.3   T. 4

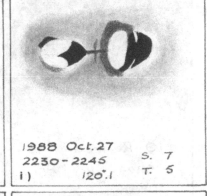

1988 Oct. 27
2230 - 2245   S. 7
i)   120.1   T. 5

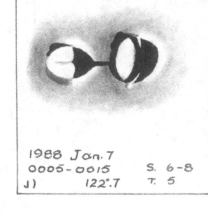

1988 Jan. 7
0005 - 0015   S. 6-8
J)   122.7   T. 5

1988 Nov. 26
2215 - 2235   S 8
k)   125.1   T. 3

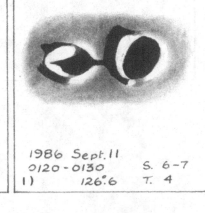

1986 Sept. 11
0120 - 0130   S. 6-7
l)   126.6   T. 4

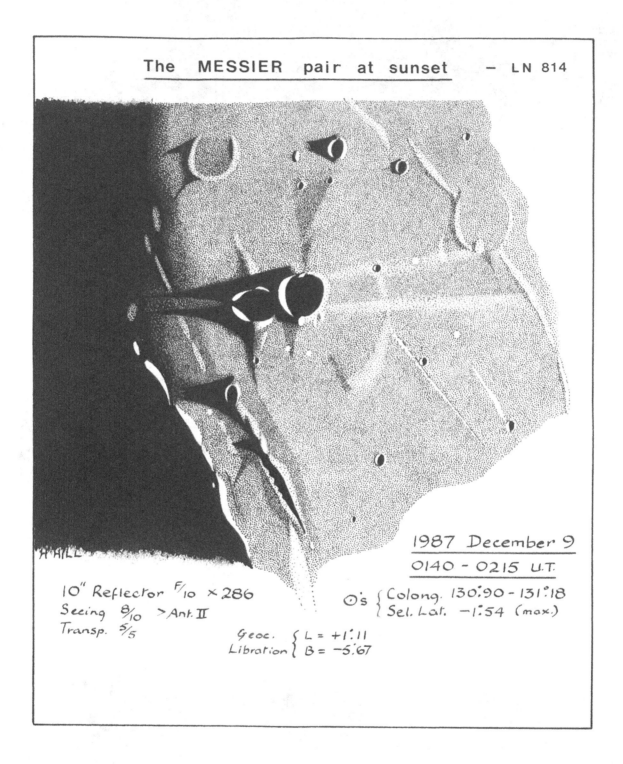

The   MESSIER   pair   at   sunset   — LN 814

H HILL

1987 December 9

0140 - 0215 U.T.

10" Reflector $^F/_{10}$ × 286
Seeing $^8/_{10}$ > Ant. II
Transp. $^5/_5$

☉'s $\begin{cases} \text{Colong. } 130°.90 - 131°.18 \\ \text{Sel. Lat. } -1°.54 \text{ (max.)} \end{cases}$

Geoc. $\begin{cases} L = +1°.11 \\ B = -5°.67 \end{cases}$
Libration

Webb and environs

Notes This region had been earmarked for study for some time on account of the rille running northwards from Webb which consists of crater-like enlargements and runs parallel to a shallow valley to the west which embouchures into the ruined enclosure Webb R. On this occasion, however, the lighting was rather too low to show the more interesting features to advantage. Ideally, observations should be carried out between colongitude 112° and 115°.

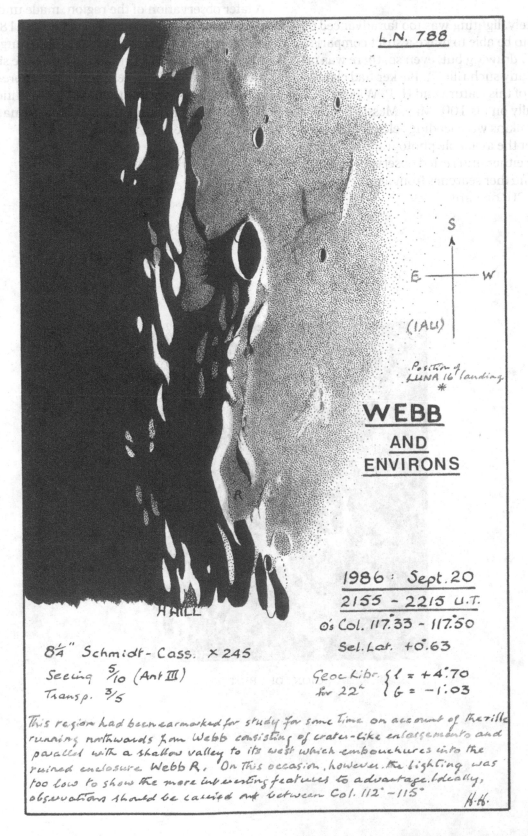

L.N. 788

S
E — W
(IAU)

Position of LUNA 16 landing
*

WEBB
AND
ENVIRONS

1986: Sept. 20
2155 – 2215 U.T.
O's Col. 117°.33 – 117°.50
Sel. Lat. +0°.63

Geoc. Libr. $\{ l = +4°.70$
for 22^h $\{ b = -1°.03$

8¼" Schmidt-Cass. ×245
Seeing ⁵/₁₀ (Ant III)
Transp. ³/₅

This region had been earmarked for study for some time on account of the rille running northwards from Webb consisting of crater-like enlargements and parallel with a shallow valley to its west which embouchures into the ruined enclosure Webb R. On this occasion, however, the lighting was too low to show the more interesting features to advantage. Ideally, observations should be carried out between Col. 112° – 115°

H.H.

The Biot Region

The observation of 1986 August 23 (opposite) was made in an attempt to confirm or refute the existence of the delicate straight rille depicted by L. F. Ball on 1933 August 8 as running from the S wall of Biot in a SE direction to the outer slopes of Wrottesley.

Unfortunately, lighting was too far advanced on this occasion to be able to make a strict comparison with the 1933 drawing but, even so, there was no suggestion of any such rille. R. Barker had claimed confirmation of this feature and H. P. Wilkins entered it boldly on his 100″ Map. Messrs Ball, Barker and Wilkins were leading selenographers of that period yet the available photographs do not show the rille either. Interested observers might care to make further searches from about colongitude 120° onwards.

A facsimile of L. F. Ball's original drawing is shown below and is taken from the *B.A.A. Journal*, **44**, No. 8. I am grateful to Mr Ball and to the Association for permission to reproduce this here.

A later observation of the region, made under perfect seeing conditions appears on p. 218. The evidence from this suggests that the arrangement of the bright interiors of craterlets along the site of the alleged rille *may* have resulted in the impression of a linear rille under less than perfect conditions. This would appear to be the most likely explanation but may not be the final word!

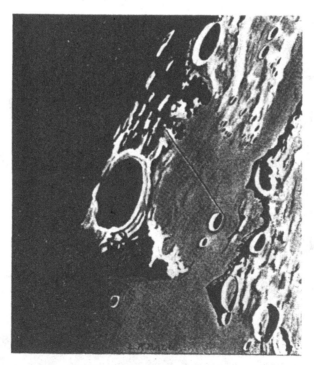

REGION OF BIOT

LN 787

BIOT

A

C

BIOT

H HILL

BIOT B

THE
BIOT REGION

1986 August 23
0305 - 0350 U.T.
O's Col. 125°.88 - 126°.30
Sel. Lat. +1°.25

8¼" Schmidt-Cass. ×245
Seeing 5-6/10 > Ant. III
Transp. 4/5
{ξ = +6°.03
{b = +1°.20

THE CRATER **BIOT**
AND ENVIRONS

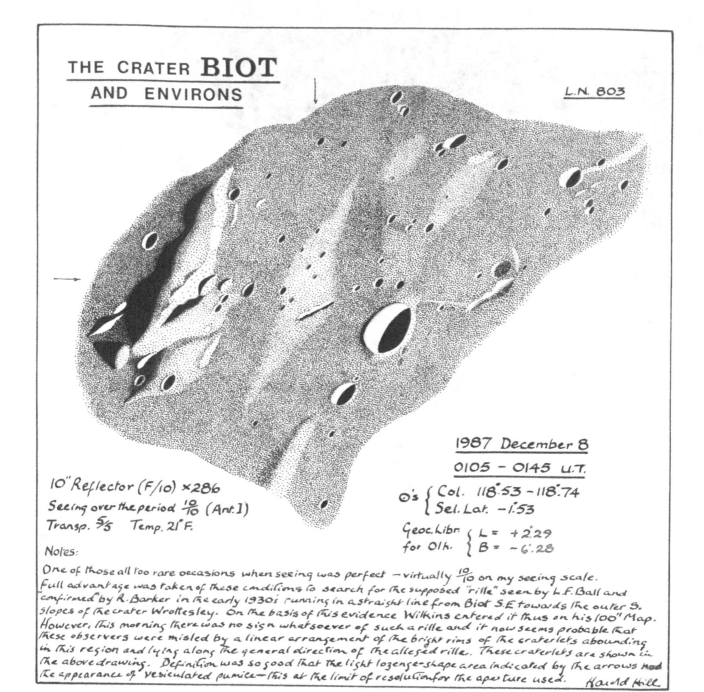

L.N. 803

1987 December 8

0105 – 0145 U.T.

⊙'s { Col. 118°.53 – 118°.74
     { Sel. Lat. – 1°.53

Geoc. Libn. { L = + 2°.29
for 01h.   { B = – 6°.28

10" Reflector (F/10) ×286
Seeing over the period 10/10 (Ant. 1)
Transp. 5/5   Temp. 21° F.

Notes:

One of those all too rare occasions when seeing was perfect – virtually 10/10 on my seeing scale.
Full advantage was taken of these conditions to search for the supposed "rille" seen by L.F. Ball and
"confirmed" by R. Barker in the early 1930's running in a straight line from Biot S.E. towards the outer S.
slopes of the crater Wrottesley. On the basis of this evidence Wilkins entered it thus on his 100" Map.
However, this morning there was no sign whatsoever of such a rille and it now seems probable that
these observers were misled by a linear arrangement of the bright rims of the craterlets abounding
in this region and lying along the general direction of the alleged rille. These craterlets are shown in
the above drawing. Definition was so good that the light lozenge-shape area indicated by the arrows had
the appearance of vesiculated pumice – this at the limit of resolution for the aperture used.   Harold Hill

Vendelinus

As a telescopic object, Vendelinus is not to be compared perhaps with its magnificent neighbours Langrenus and Petavius, as it is in a comparatively ruinous condition with broken and irregular ramparts, but, despite this, it abounds in many interesting features both in its interior and on the surrounding walls so that it is a most attractive spectacle when seen under late afternoon or evening light. Equally, the last shafts of sunlight to be thrown across its floor as sunset approaches form an absorbing sequence for study if followed through to extinction.

Chosen from a number of observations made at various times in the 1980s, the drawing opposite shows the region at what is probably the best phase for interior details before the shadows from the west wall become obtrusive.

Many anomalies from the past continue to offer challenges to new generations of observers. Outstanding among these must be the Vendelinus rille seen in 1881 by Elger, and in the following year by Maw whose drawings and account of the rille are quite unequivocal. It also appears in the lunar maps of both Goodacre and Wilkins as an unmistakeable feature running northwards from the small floor crater Vendelinus H. This rille was sought for in vain by observers of repute such as Barker and Moore; even Wilkins was later to deny its existence mainly as a result of later observations with larger apertures. The whole matter was resurrected when the disputed rille was recovered in part by Thornton (18″ refl.) and Haas (18″ refr.) and also by other observers such as Cooke who saw the rille essentially as drawn by Elger – though issuing from the twin features NW of H rather than the craterlet itself.

Vend. H is indicated on the drawing shown here and though the *position* of the alleged rille follows the line of demarcation between two areas of different tone (change of slope?) I have never seen any sign of the rille at favourable opportunities, despite the utmost attention. Where the rille is portrayed by both Elger and Cooke as passing through the crater L further north, I could only make out a coarse, shaded valley here – the latter with a bright border in its northern section suggestive of a graben feature, but no rille as such.

The large formation Lamé which is shadow-filled in the drawing and encroaches on the NE side of Vendelinus is bordered on its outer W ramparts by two large semi-circular formations which extend on to the floor of Vendelinus as either remnants of old rings or perhaps landslips. These are fascinating subjects for observation as they contain 'test objects' on their inner and outer slopes in the form of craterlets and craterpits – these are seen to greater advantage at a somewhat earlier stage of lighting than that shown. On the floor of Vendelinus itself, south of these 'landslips' I caught several craterlets not previously recorded, three or four of which bore signs of being crater-cones – the lunar equivalent of our volcanoes? They were evanescent as shortly afterwards their shaded eastern slopes were lost in the general darkening of the floor hereabouts. The horseshoe ring to the NW of these was suggestive of a partially submerged formation.

There is still much to engage the interest of the telescopist in the region as a whole.

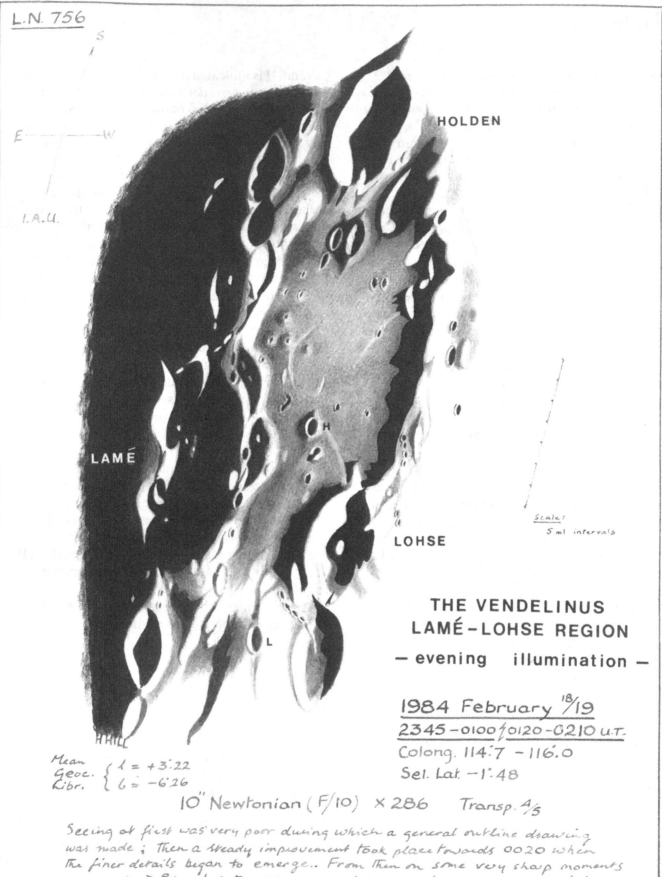

L.N. 756

S

E —— W

I.A.U.

HOLDEN

LAMÉ

H

LOHSE

L

H HILL

Scale:
5 ml intervals

**THE VENDELINUS
LAMÉ–LOHSE REGION**

– evening   illumination –

1984 February 18/19
2345–0100 / 0120–0210 u.t.
Colong. 114°.7 – 116°.0
Sel. Lat. –1°.48

Mean
Geoc. { ℓ = +3°.22
Libr. { b = –6°.26

10" Newtonian (F/10) × 286   Transp. 4/5

Seeing at first was very poor during which a general outline drawing
was made ; Then a steady improvement took place towards 0020 when
the finer details began to emerge.. From then on some very sharp moments
occurred > 8/10 but the general quality was between 6 & 7. I have
never recorded so much within & around Vendelinus before.
Discussion of observational features is to be found elsewhere.

Harold Hill

220

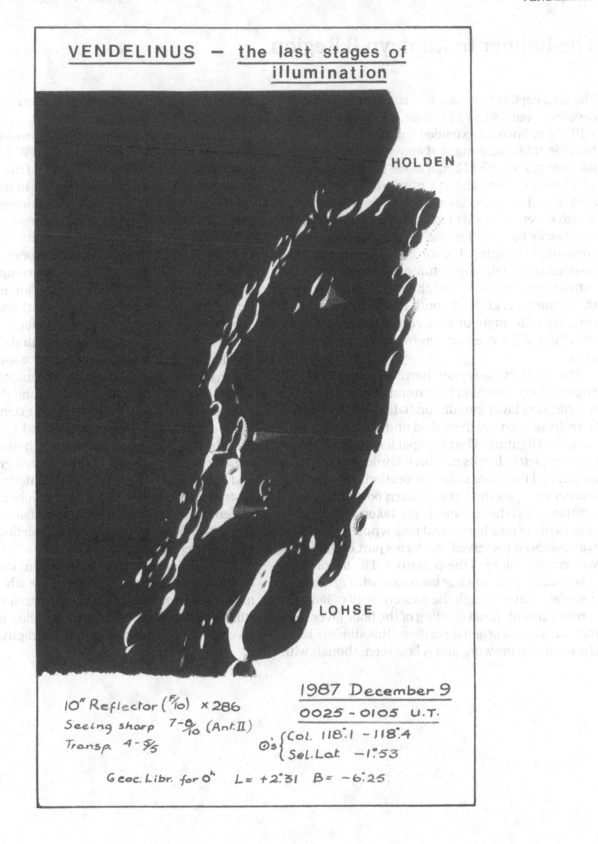

The Balmer to Kapteyn B Region

The area depicted lies east and south of Vendelinus covering from 14° S to 26°S lat. – a distance of over 220 miles. Such an extended coverage could hardly be undertaken on a night of above-average seeing because the wealth of detail alone would preclude effective plotting of all but a much more limited area. It will be appreciated that an observation which covers a period of nearly three hours will necessarily be 'out-of-phase' as the aspect is constantly changing. It is a composite of three sessions at the telescope running almost concurrently on the one night and working down the terminator, i.e. from south to north. Seeing conditions throughout were rather indifferent and the object of the exercise was to obtain a general view.

The *details* of the region, however, have been engaging my attention for a number of years and the effects of laval inundation, which are so evident here, have been much studied under different aspects of lighting. The principal formation, Balmer, has completely lost its northern border from this cause, and there are strong indications that the true Balmer occupies the larger eastern portion of the southern bay, the remains being taken up by the rudiments of another eroded ring whose western wall has been preserved and forms part of the western boundary of the enclosure. The flooding of this southern portion has been extensive and almost complete, but although the western wall of Balmer is non-existent, tonal shading of the floor gives some indication of its original position; this shading is not shown in the drawing and is best seen, though with

difficulty, under morning illumination as soon as early shadow has lifted.

A much clearer example of laval erosion is to be found in the confluent rings Phillips W & D further to the south. Here the eastern wall of the overlapping D is very low but visible in its entirety under oblique lighting and it is emphasised by the greater relative darkness of W's floor as sunset approaches. At this time also, the eastern rampart of W, though fragmented is quite prominent but the northern border of this formation is completely breached, suggesting that the inundation came from the north–especially as the southernmost rims of both W & D remain relatively intact.

The composite drawing suggests that the eroded Balmer and its companion occupy the southern end of a marial tract or, perhaps, rudimentary walled-plain about 195 miles in diameter (and therefore comparable to Bailly in size), which extends almost as far north as the crater Kapteyn at 11° S 71° E (not shown). The expanse includes, however, the more prominent objects of the Kapteyn group, A and B, whilst the remainder of the interior is bestrewn with minor elevations, hillocks, ridges, craterlets, etc; it is not a recognised feature on the maps – presumably because of its indefinite borders – and is therefore unnamed.

It is hoped, in due course, to present the detailed findings of several years' study of the whole region in treatise form. This plain is a vast area which will still repay exploration, notwithstanding the 'overhead' results of Orbiter photography.

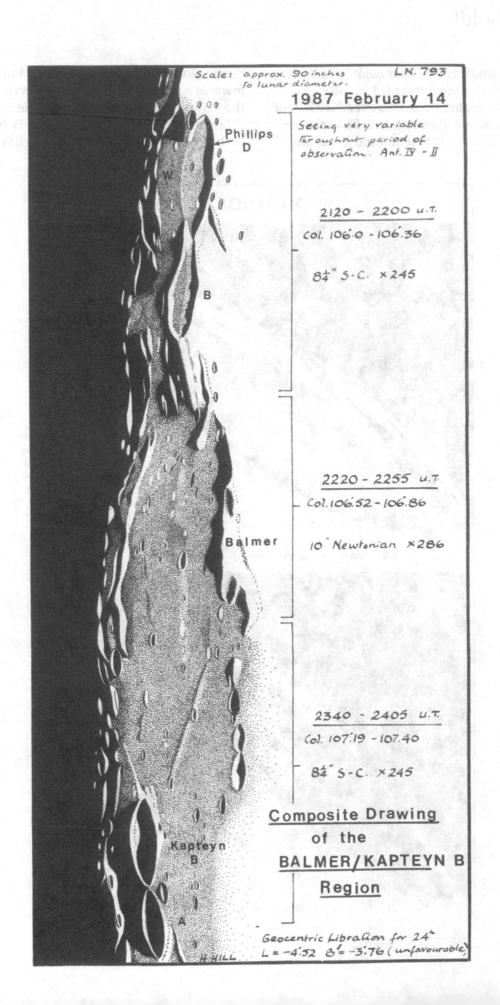

Scale: approx. 90 inches to lunar diameter.

L.N. 793

1987 February 14

Seeing very variable throughout period of observation. Ant. IV – II

2120 – 2200 U.T.

Col. 106.0 – 106.36

8¼" S-C. ×245

2220 – 2255 U.T.

Col. 106.52 – 106.86

10" Newtonian ×286

2340 – 2405 U.T.

Col. 107.19 – 107.40

8¼" S-C. ×245

Composite Drawing
of the
BALMER/KAPTEYN B
Region

Geocentric Libration for 24ʰ
L = –4°.52 B' = –3°.76 (unfavourable)

Phillips D

W

B

Balmer

Kapteyn B

A

P. HILL

223

W. Humboldt

Three evening studies of W. Humboldt have been selected to show this magnificent, 125 miles diameter, walled-plain under different conditions of both illumination and libration. Its lofty walls rise in majestic peaks to 16 000 ft in some places and the massive eastern ramparts are a fine spectacle towards sunset as they are deeply terraced, part of the NE border being distinctly double.

Libratory conditions were particularly favourable in the first drawing, and with the lighting around

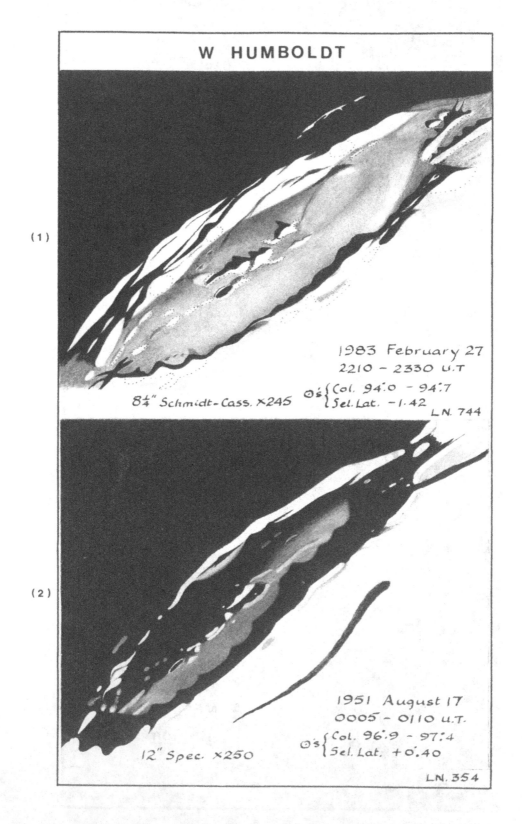

224

colongitude 94°, the details of the floor consisting of a central mountain group, craters and lesser elevations are all displayed to best advantage. In a matter of a few hours, large areas of the floor are covered in shadow; these develop not only from the western wall but also the eastern half of the interior which darkens as a consequence of surface curvature – particularly manifest in formations of large dimensions. This is well shown in drawing (2).

Even under optimum presentation, W. Humboldt

is still too near to the limb for effective observation of the remarkably complex system of radial and concentric rilles which Orbiter photography has revealed on its interior, although glimpses have been obtained of two or three of the larger members whose direction lies approximately along the observer's line of sight . . . see drawing (3) below in which two rilles were suspected and one across the southern floor which was held with certainty.

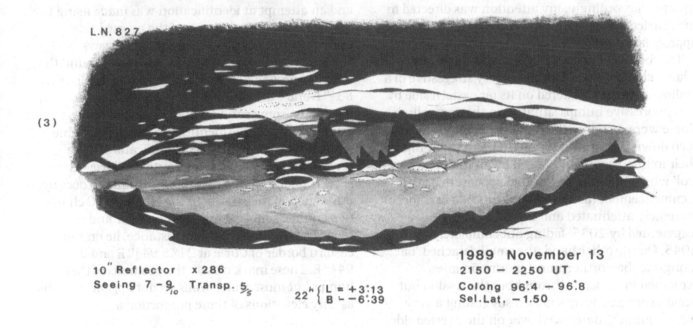

L.N. 827

(3)

10″ Reflector x 286

Seeing 7 - 9/10 Transp. 5/5

22ʰ { L = +3°.13 B = -6°.39

1989 November 13

2150 — 2250 UT

Colong 96°.4 — 96°.8

Sel. Lat, -1.50

Observations of the averted hemisphere

The notes which deal with Neper and the Mare Smythii (see Quad.I Sect.3) refer to the exceptional conditions under which these areas were observed to advantage in the Saros cycles 1950, 1968 and 1987. Similar circumstances occurred in the late months of 1988 for a different portion of the limb, namely, the eastern section lying *south* of the lunar equator; accordingly, my attention was directed to the districts lying beyond W. Humboldt on the appropriate dates.

The footnotes to the drawing (opposite) refer to a large shadow-filled formation very suggestive of a walled-plain and bordered on its far eastern side by two impressive humps situated on the actual limb. These were certainly the highest elevations to be seen down the terminator on that particular night; their loftiness was emphasised by an intervening 'col' which presented a knotted appearance at commencement (as shown) but this had become extremely attenuated and then delicately fragmented by 2035, fading out completely by 2045. On the other hand, the small detached 'bar' alongside the northernmost mountain mass remained in evidence throughout the session but became progressively smaller, suggesting a very high summit situated well over on the averted side.

In seeking to identify features in limb zones it is necessary to plot with care the brighter and more obvious foreground objects whose positions are known or can be readily determined from maps/ photographs, etc. In the present case, consultation of all available material failed to reveal any previous observational record of this particular zonal strip and an attempt at identification was made using the NASA Aeronautical Charts of the averted side drawn up from the results of Orbiter surveys. Having regard to position and size, it was found that the most probable formations were Gibbs at $18\frac{1}{2}°$ S $83\frac{1}{2}°$ E and Curie at $23\frac{1}{2}°$ S 92° E although, inevitably, some residual doubts exist over establishing positive identities at this stage. Some ambiguity exists regarding identification of the prominent limb crater lying south of 'Curie' as several undesignated objects were found to occupy this region of the charts. In the NASA ACIC charts two craters, some 30 miles in diameter and separated by about half that distance, lie on the eastern border of Curie at 21° S $94\frac{1}{4}°$ E and 24° S $94\frac{1}{2}°$ E. These mark closely the positions of these humps but most crater walls do not appear in profile as lofty elevations of these proportions.

226

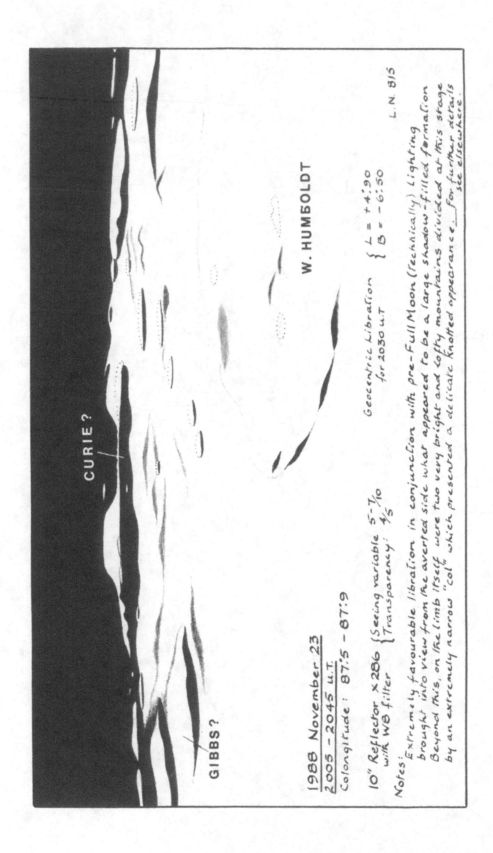

GIBBS ?

CURIE ?

W. HUMBOLDT

1988 November 23
2005 - 2045 u.t.
Colongitude : 87.°5 - 87.°9

10" Reflector ×286 { Seeing variable 5-7/10
with W8 filter { Transparency: 4/5

Geocentric Libration { L = +4.°90
for 2030 u.t { B = - 6.°50

L.N. 815

Notes: Extremely favourable libration in conjunction with pre-Full Moon (technically) lighting
brought into view from the averted side what appeared to be a large shadow-filled formation
Beyond this, on the limb itself were two very bright and lofty mountains divided at this stage
by an extremely narrow "col" which presented a delicate knotted appearance. For further details
see elsewhere.

Quadrant IV – Section 16

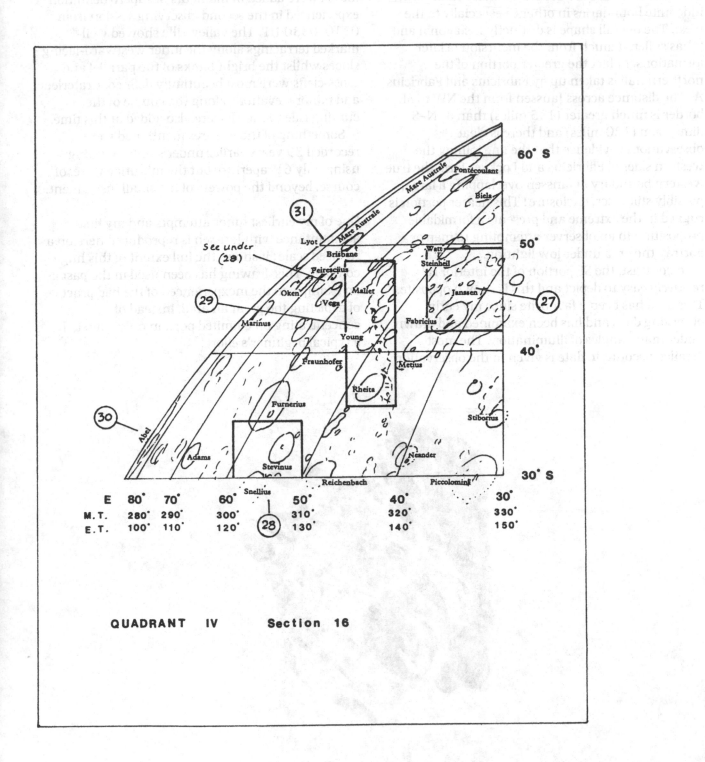

QUADRANT IV   Section 16

Janssen

This enormous mountain-walled enclosure has every appearance of being one of the oldest formations on the visible hemisphere. The surrounding walls are broken in many places with indefinite boundaries in others – especially to the west. The overall shape is distinctly hexagonal and it has suffered much from the incursion of later formations, in fact the greater portion of the northern half is taken up by Fabricius and Fabricius A. The distance across Janssen from the NW to SE border is much greater (155 miles) than its N–S dimension (120 miles) and there is clear observational evidence that the line joining the western sides of Fabricius and Lockyer form the true western boundary of Janssen overlapping a large, possibly still older, enclosure! This nearer portion is rugged in the extreme and presents a formidable proposition to an observer attempting to map or portray the area under low lighting.

In contrast, the SE portion of the interior is *relatively* easy to depict and this is shown opposite. This area has been a favourite since my early observing days and has been examined and drawn under many angles of illumination. The most detailed account to date is given in the observation of 1987 Sept.10/11. The finer details of the great valley-rille and its off-branches to the south, together with the features of the complex central massif were added in moments of superb definition experienced in the second observing session from 0310–0330 UT. The valley-rille showed well-marked terracings along the inner westward-facing slopes whilst the bright banks of the parallel and cross-clefts were most beautifully defined. Craterlets and minor elevations along the course of the curving rille to the SE were also added at this time.

Something of the features mentioned were recorded 39 years earlier under similar lighting using only $6\frac{1}{2}''$ aperture but the minutiae were, of course, beyond the powers of the small instrument.

One of my earliest lunar attempts and my first acquaintance with Janssen is reproduced here on a reduced scale; it shows the full extent of this huge complex. This drawing has been used in the past as an example to the inexperienced of the bad practice of depicting too great an area, instead of concentrating on a limited portion only – as such, a typical beginner's effort!

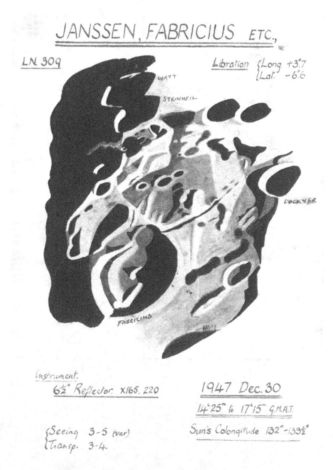

JANSSEN, FABRICIUS ETC.,

LN. 309

Libration { Long. +3°.7
{ Lat. −6°.6

WATT

STEINHEIL

LOCKYER

FABRICIUS

HILL

Instrument.
$6\frac{1}{2}''$ Reflector. X165. 220

1947 Dec. 30
14ʰ25ᵐ to 17ʰ15ᵐ G.M.A.T.

Sun's Colongitude 132°–133½°

{ Seeing 3-5 (var)
{ Transp. 3-4.

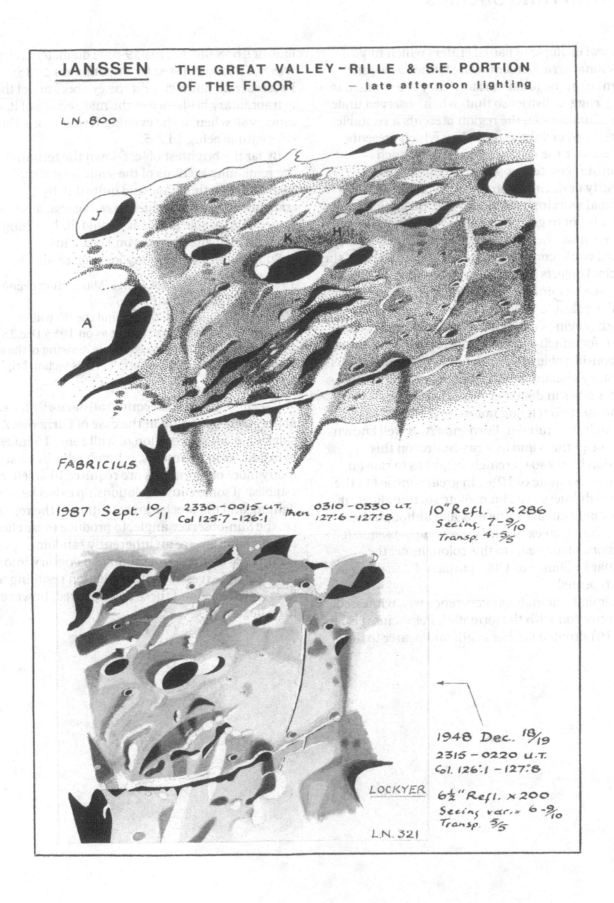

JANSSEN — THE GREAT VALLEY–RILLE & S.E. PORTION OF THE FLOOR — late afternoon lighting

L.N. 800

FABRICIUS

1987 Sept. 10/11   2330–0015 u.t.   then   0310–0330 u.t.
Col 125°.7 – 126°.1   127°.6 – 127°.8

10" Refl.   ×286
Seeing 7–9/10
Transp. 4–5/5

1948 Dec. 18/19
2315 – 0220 u.t.
Col. 126°.1 – 127°.8

LOCKYER

6½" Refl. ×200
Seeing var.= 6–9/10
Transp. 5/5

L.N. 321

Stevinus and Snellius

A typical example of paired craters which have a meridional arrangement – one found to be quite common on the lunar surface. They are situated in a very rugged district so that, when observed under a low illumination, the region presents a veritable chiaroscuro of light and shadow which presents problems for the draughtsman. Under such circumstances, telescopic drive constitutes a necessity as distinct from a luxury, as it facilitates the rapid working which is so essential if the detail drawn is not to get 'out-of-phase'. It will be noted that, even so, the observation took all of $1\frac{1}{2}$ hours of hard work, commencing with the outlining of the principal objects and then working northwards down the terminator before much of this transient detail was lost to view.

Both Stevinus and Snellius are imposing craters with lofty, much-terraced walls, central elevations, and considerable interior detail which, however, is hidden by shadow at this stage; the craters are 46 and 52 miles in diameter respectively.

The inset sketch and notes are pertinent as although the craterlet, Furnerius A, is well known for its brightness and as a ray centre, on this occasion it was so piercingly bright as to rate an intensity estimate of $10+$. In great contrast to the general dullness of its immediate surroundings, it continually attracted the eye, and although the region has been examined on several subsequent occasions at, or near, to this colongitude, the singular brilliance of 1983 January 1 has never been repeated.

A similar anomalous occurrence was witnessed in connection with the formation Peirescius (also in Sect.16) situated further south and nearer to the limb at 46° S 68° E. This 39-mile diameter crater was not a subject listed for attention on 1985 December 28 but it caught the eye because of the extraordinary brilliance of the mid-section of its east inner wall when at the evening terminator – the colongitude being 112°5.

By far the brightest object down the terminator, the remaining sections of the same wall were fragmented at this stage and quite dull by comparison. It was decided to re-observe at some future comparable phase as I could find nothing relevant in available notes on Peirescius.

Subsequent entries in my logbook read:

1986 Feb.26 0145 UT col.112°4 'Mid-section definitely not as bright as two months ago'.
1986 Oct.20 col.111°9 'Do not find the W. wall of Peirescius anything like so bright as on 1985 Dec.28'.
1989 Jan.23 col.111°2 – a complete drawing of the area was made with the annotation: '. . . this section bright but *not* brilliant'.

It seems difficult to account satisfactorily for such abnormal brightening in the case of Furnerius A – still less so where a section of wall some 15 miles in extent, as in Peirescius, is involved! Obviously, many more observations are required in an effort to establish if some causal relationship exists between lighting and the precise presentation of the region, i.e. the Sun-observer angle, to produce these effects or whether they are an inherently random occurrence of unknown origin. An enquiry into the possibility of excessive solar radiation resulting from flare activity around these dates proved, however, to be negative.

STEVINUS and SNELLIUS
at the evening terminator

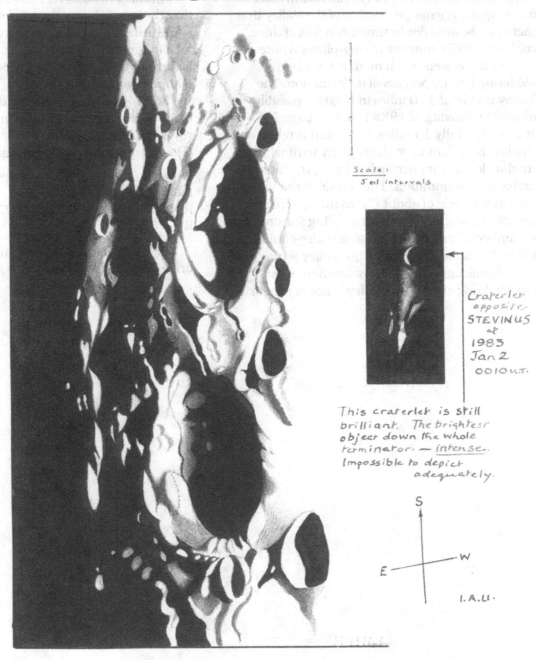

Scale:
5 ml intervals

Craterlet
opposite
STEVINUS
at
1983
Jan 2
0010 u.t.

This craterlet is still
brilliant. The brightest
object down the whole
terminator. — intense.
Impossible to depict
adequately.

S

E — W

I.A.U.

1983 JAN. 1
2225 - 2355 U.T.

Geoc. Libr. { Long +3°.80
{ Lat. -4°.55

Sun's Col. 120°.8 - 121°.5

203mm. Schmidt-Cass. ×245   Seeing 7/10 Transp 4/5

The Great Rheita Valley

This is the largest and most important of its class on the part of the lunar surface within reach of the terrestrial observer. So many crateriform objects make up the greater proportion of the Valley that there can be little doubt that it consists of the coalescence of a number of ring-plains whose borders have been much modified or even obliterated in the process of its formation. The Valley is at least 120 miles in length – possibly more, see drawing of 1988 Feb.5 on page 237 – and at its widest fully 15 miles, but that it is relatively shallow is evident from the rapidity with which shadow leaves (or encroaches upon) the interior under low illumination. The 'strike' of the Valley runs at an angle of about 60° to the apparent meridional alignment of surrounding features, but a number of crater-couplets and crater-chains follow the same direction as the Valley which may be of significance in any consideration of its morphology; clearly the Valley is not of meteoric origin though any discussion of theories of its origin is outside the scope of this portfolio.

A number of observers have claimed in the past that the inner slopes of the formation Young have a greenish, almost translucent cast or sheen when seen at the evening terminator. The writer has looked for this effect on many occasions but always without success.

The two morning studies of the Rheita Valley made on successive evenings in 1989 February show the change of aspect over 24 hours. The relevant observational details are:

(a) Feb.9 1710–1825 UT Col.$314°.7–315°.3$ L$= +2°.40$ B$= -3°.69$
(b) Feb.10 1730–1840 UT Col.$327°.0–327°.6$ L$= +3°.76$ B$= -5°.00$

The $8\frac{1}{4}''$ Schmidt-Cass. was used at $\times 245$ with drive on both occasions.

234

LN 818

TWO STUDIES OF THE GREAT

RHEITA VALLEY

a) 1989: Feb 9

&

b) 1989: Feb 10

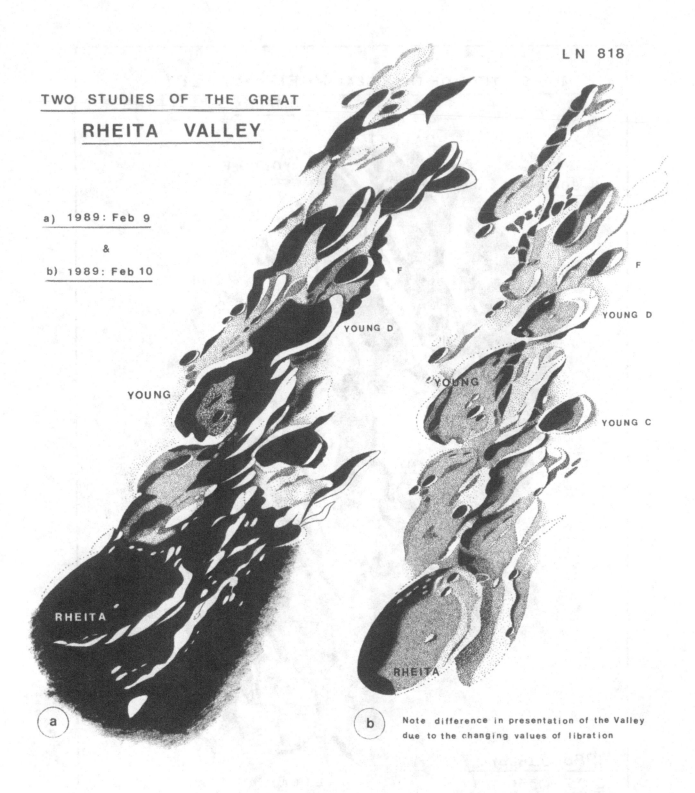

F

YOUNG D

YOUNG

YOUNG C

F

YOUNG D

YOUNG

RHEITA

RHEITA

a

b

Note difference in presentation of the Valley
due to the changing values of libration

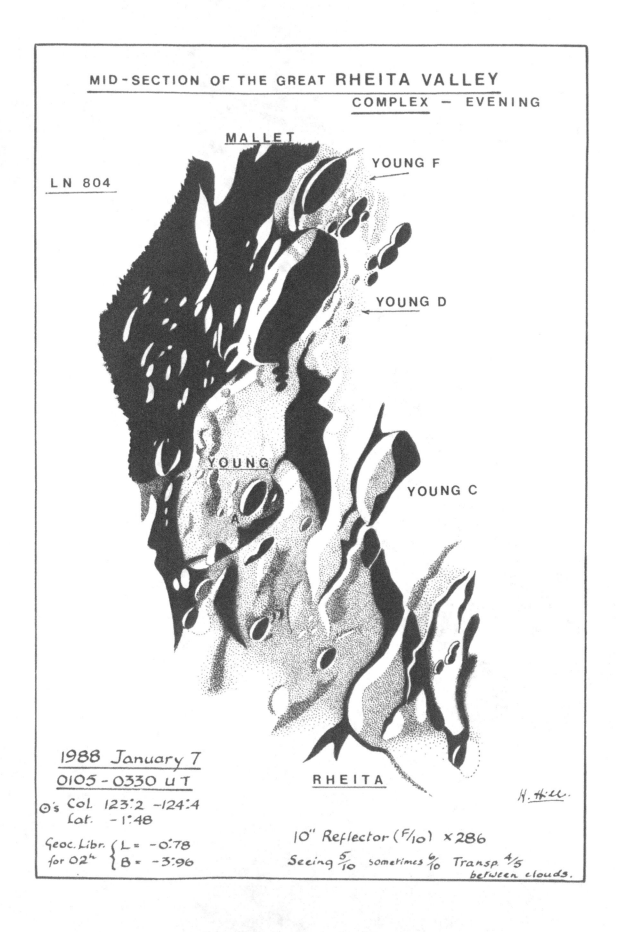

MID-SECTION OF THE GREAT RHEITA VALLEY
COMPLEX — EVENING

MALLET

YOUNG F

LN 804

YOUNG D

YOUNG

YOUNG C

A

RHEITA

1988 January 7

0105 - 0330 UT

H. Hill

⊙'s Col. 123°.2 -124°.4
    Lat. -1°.48

Geoc. Libr. { L = -0°.78
for 02ʰ   { B = -3°.96

10" Reflector (F/10) ×286

Seeing 5/10 sometimes 6/10 Transp. 4/5
between clouds.

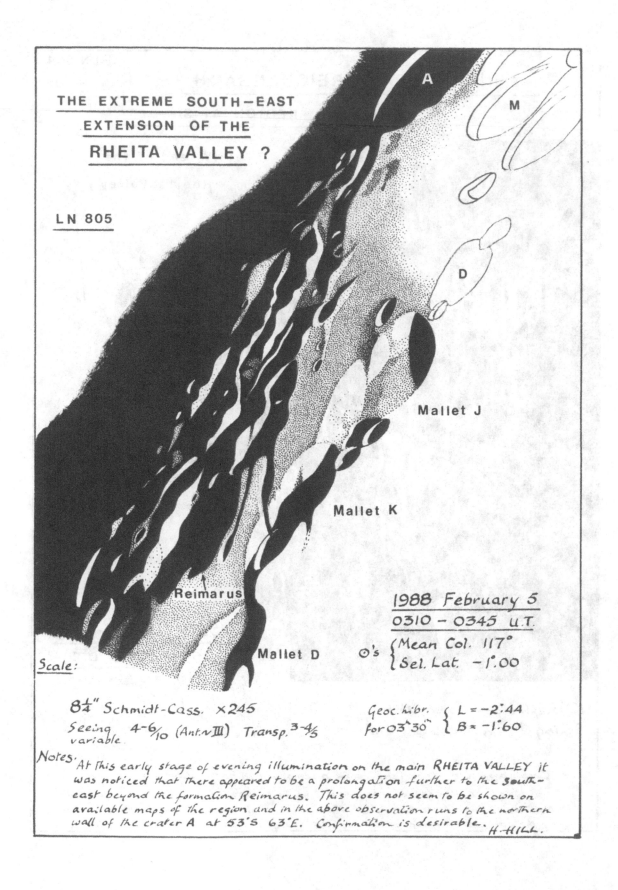

THE EXTREME SOUTH—EAST
EXTENSION OF THE
RHEITA VALLEY ?

LN 805

A

M

D

Mallet J

Mallet K

Reimarus

Mallet D

Scale:

1988 February 5
0310 – 0345 U.T.

☉'s { Mean Col. 117°
      { Sel. Lat. –1°.00

8¼" Schmidt-Cass. ×245
Seeing 4-6/10 (Ant.∼III) Transp. 3-4/5 variable.

Geoc. libr. { L = –2°44
for 03ʰ30ᵐ { B = –1°60

Notes: At this early stage of evening illumination on the main RHEITA VALLEY it was noticed that there appeared to be a prolongation further to the south-east beyond the formation Reimarus. This does not seem to be shown on available maps of the region and in the above observation runs to the northern wall of the crater A at 53°S 63°E. Confirmation is desirable. H. Hill.

237

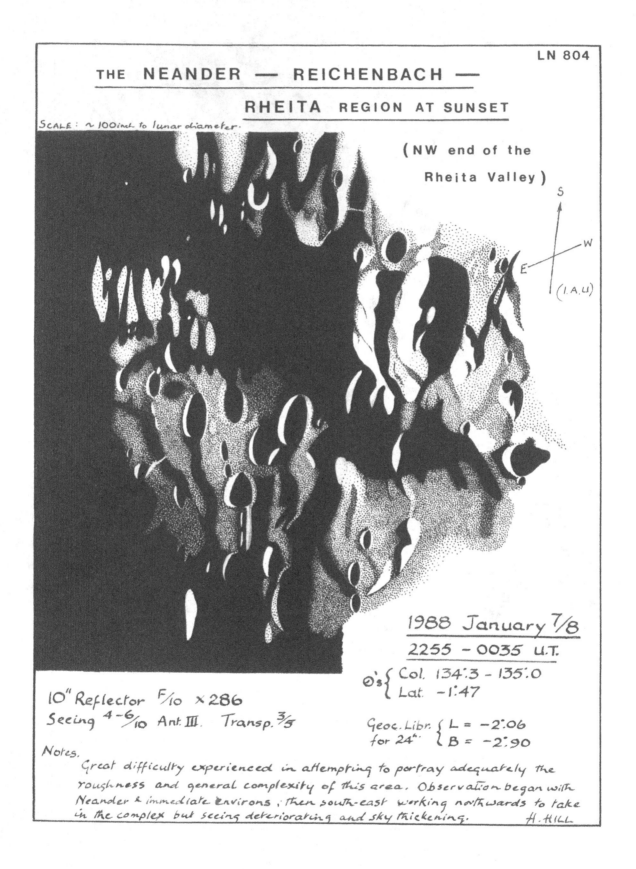

LN 804

THE NEANDER — REICHENBACH —
RHEITA REGION AT SUNSET

SCALE : ~ 100 inch to lunar diameter.

(NW end of the
Rheita Valley)

1988 January 7/8

2255 – 0035 U.T.

☉'s { Col. 134°.3 – 135°.0
      Lat. –1°.47

Geoc. Libr. { L = –2°.06
for 24ʰ   { B = –2°.90

10" Reflector F/10 ×286
Seeing 4-6/10 Ant. III. Transp. 3/5

Notes.
   Great difficulty experienced in attempting to portray adequately the
roughness and general complexity of this area. Observation began with
Neander & immediate environs; then south-east working northwards to take
in the complex but seeing deteriorating and sky thickening.   H. Hill

Limb regions south and east of W. Humboldt

The mean position of W. Humboldt is 27° S 81° E which means that to examine the regions in the libratory zone to the south and east of this object circumstances must be favourable and the timing critical. Although in the present instance the drawings were made at what was *technically* the Full Moon phase, the defect of illumination along the east limb caused by a suitable combination of libration in longitude and, *importantly*, latitude, showed the formations Barnard and Abel to advantage. Both are walled-plains of the order of 75 miles in diameter and representative of their class

in having smooth, darkish floors which contain little detectable detail – though seeing was far from good on this occasion and the statement is open to revision. Barnard and Abel are situated at 29° S 86° E and 34° S 85° E respectively, so that under mean libration they are positioned almost on the limb, and situated beyond it under adverse conditions.

These formations will be presented under virtually optimum conditions in the year 2001 when the Saros cycle of 18 years' duration favours the SE limb.

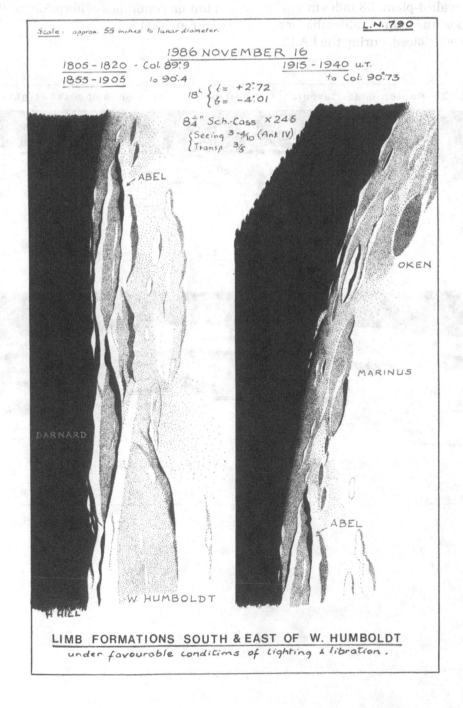

LIMB FORMATIONS SOUTH & EAST OF W. HUMBOLDT
under favourable conditions of lighting & libration.

The Lyot Region

The Mare Australe (the Southern Sea) is something of a misnomer since it consists largely of a collection of ring plains most of which have darkish interiors, and these are interspersed with dark irregular patches of a similar grey hue, though varying somewhat in intensity, but nowhere is there a continuous sheet of dark material which characterises the Maria proper. The much foreshortened view we get from Earth must have given rise to this impression, hence the appellation.

Prominent among the dusky enclosures in the Australe area are the 50 mile diameter Oken and the larger flooded walled-plain, 88 miles in diameter, which held the name Pratdesabà for a time before this was replaced, during the I.A.U.

Commission's revision of lunar nomenclature, by Lyot – formerly attached to Ptolemaeus A.

The libratory swing favoured the SE limb during times of appropriate lighting in the early months of 1989 and the opportunities were taken to observe the Lyot region under both morning and evening light and the results are shown opposite. Little is shown, however, which was not recorded in the early work of Abineri, Lenham and Moore who were pioneers in plotting this region during the 1950s. Nevertheless, the area is attractive enough to be commended for further examination; optimum conditions of libration for this limb will occur in 2001.

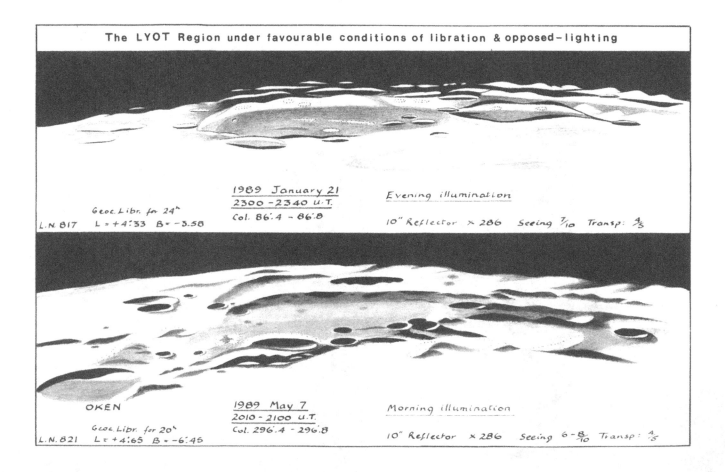

The LYOT Region under favourable conditions of libration & opposed-lighting

1989 January 21
2300 – 2340 U.T.
Col. 86°.4 – 86°.8

Evening illumination

L.N. 817   Geoc. Libr. for 24ʰ   L = +4°.33   B = –3.58

10" Reflector × 286   Seeing ⁷⁄₁₀   Transp: ⁴⁄₅

OKEN

1989 May 7
2010 – 2100 U.T.
Col. 296°.4 – 296°.8

Morning illumination

L.N. 821   Geoc. Libr. for 20ʰ   L = +4°.65   B = –6°.45

10" Reflector × 286   Seeing 6 – 8⁄10   Transp: ⁴⁄₅

Printed in the United States
By Bookmasters